Environmental Philosophy, Politics, and Policy

Environmental Philosophy, Politics, and Policy

Edited by
John A. Duerk

LEXINGTON BOOKS
Lanham • Boulder • New York • London

Published by Lexington Books
An imprint of The Rowman & Littlefield Publishing Group, Inc.
4501 Forbes Boulevard, Suite 200, Lanham, Maryland 20706
www.rowman.com

6 Tinworth Street, London SE11 5AL, United Kingdom

British Library Cataloguing in Publication Information Available

Library of Congress Cataloging-in-Publication Data

Library of Congress Control Number: 2020949609

ISBN 978-1-7936-1763-7 (cloth)
ISBN 978-1-7936-1765-1 (pbk)
ISBN 978-1-7936-1764-4 (electronic)

Contents

List of Illustrations

FIGURES

TABLES

Introduction

John A. Duerk

Somewhere around my graduation from high school, I started collecting newspapers, plastic containers, glass jars, and aluminum cans at home because discarding these items with the other refuse made me uncomfortable. Unfortunately, the city where I lived did not yet offer curbside recycling in my neighborhood. Thankfully, my siblings and parents joined me in this effort. Every couple of weeks I would drive what we collected to a nearby village that had a recycling drop-off site, which consisted of multiple industrial dumpsters. Then, later that fall, I wrote my first letter to the editor of the local newspaper about reconsidering what we dispose in landfills. These small experiences served as a foundation to my environmental ethos and a shift in my thinking as a young adult. They can be traced to this anthology.

Many people in the United States and around the world share a genuine concern about our impact on the land, water, air, plants, animals, and intricate ecosystems. When you look at recent Gallup poll survey data, people in the United States are mindful about some of the choices they make in their daily lives. For example, 86 percent of respondents "voluntarily recycled newspapers, glass, aluminum, motor oil or other items" and 72 percent "used reusable shopping bags at the grocery store instead of the standard plastic or paper bags" (Brenan, 2020). This is evidence that some ideas about the environment have permeated the culture. Undoubtedly, people have changed some of their regular habits. While these are notable developments, the progress some hope for is yet to be achieved. In 2017, the United States only recycled 8.4 percent of its plastic waste and then buried approximately three-fourths of it in landfills (U.S. Environmental Protection Agency, 2019). Moreover, the total amount of plastic recycled in 2017 dropped below the figure from 2015 despite the fact that the total amount generated rose (U.S. Environmental Protection Agency, n.d.).

Other numbers give us more to think about on a political level. Survey data indicate that only 32 percent of people have "contributed money to an environmental, conservation or wildlife preservation group," 19 percent "contacted a public official about an environmental issue," and 18 percent "attended a meeting concerning the environment" (Brenan, 2020). Many appear to think about their personal choices (as evidenced above with recycling), but most do not use their public voice to engage others in the political process to change existing conditions. Additional data is mixed. In 2001, 61 percent of respondents thought global warming resulted from the way that human beings live, and it increased slightly to 64 percent almost two decades later (Saad, 2020). Furthermore, we see a partisan divide on this issue between those who self-identify as Democrats and those who self-identify as Republicans (Funk and Kennedy, 2020). At the same time, 33 percent of people claimed to be deeply concerned about global warming in 2001 and that number rose thirteen points to 46 percent over a twenty-year period (Saad, 2020).

When I decided to undertake this anthology, multiple questions entered my mind as one might expect. What do I seek to achieve? Which environmental issues should be addressed? Who would be willing to contribute a chapter? The list could go on. Like any project of its kind, a person can find her or himself paralyzed by the infinite possibilities. What I quickly learned is that you need to choose a direction and start contacting scholars who share your research interests because there is an "unfolding" process to be embarked upon as editor. Then, with the proper effort and time, a solid collection of articles could be assembled.

My utmost hope is that the ideas expressed within, and findings yielded by, the chapters in this book will advance some of the conversations people are having about the environmental challenges we face as human beings. Now, these discussions may occur in a college classroom or they might be over the family dinner table. Honestly, they can—and should—take place everywhere. Wherever the conversations occur, people's ideas must be informed by trusted sources that will serve as part of the foundation to any exchange. Now, while it is impossible for any anthology to be definitive, this collection of articles contributes to our understanding of these issues at a distinctive moment in time. It is a snapshot of what a handful of contemporary scholars think deserves attention.

Another hope for this project is that it creates an academic space for scholars who are interested in exploring environmental issues within the confines of their respective academic training. There is a difference between writing a piece to advance human understanding and writing a piece to advance a political position. Furthermore, while pure objectivity might not exist in the social sciences or humanities, that does not mean academic methodology should be abandoned in favor of polemics. Surely, there must be room in the existing discourse for individuals who employ a more dispassionate approach to their work. Would a person who does not share certain values be more, or less, inclined to listen to a scholar who makes a good faith effort to restrain her or his own political views? If you write a political diatribe, chances are you will not convince anyone other than people who are already sympathetic to the position you champion.

Before proceeding with a survey of the anthology's contents, the reader must know that it has also been inspired by Edward O. Wilson's (1998) *Consilience: The Unity of Knowledge*. One of his key ideas is that we must invite multiple disciplines into the conversation if we really seek to advance our understanding of any phenomenon. As an issue, "the environment" is so broad and multifaceted that each chapter contained herein is a separate conversation unto itself. Not only is it important to discuss a variety of the (sub-) issues (e.g., bike lanes, climate change), but also have those (sub-)issues explored by academics from multiple disciplines. Wilson (1998) argues that the hard sciences and social sciences and different disciplines within the social sciences must build relationships because they need each other. While only one of the twelve contributing authors represents the hard sciences and his piece is strictly a qualitative analysis, this collection is a conscious attempt to connect multiple disciplines within the social sciences and connect the social sciences with the humanities. To me, "environmental studies" cannot simply be composed of subjects like environmental science, biology, and climatology. It must also include anthropology, economics, geography, political science, English, philosophy, and landscape architecture and forest ecology.

This work contains three different sections: philosophy, politics, and policy. The first, philosophy, is composed of three different pieces. Mark Thorsby argues that "The Land Ethic" has been part of the conversation about human responsibility to the environment since the middle of the twentieth century; yet, many people have not incorporated it into their values and lifestyle when you consider the harm that human beings have caused to the natural world around us. As consumers, perhaps it is time we reflected upon, and changed, our behavior. Next, applying John Rawls' theory of justice, Alan Clune contends that animal liberation activists should consider which animals experience the most amount of suffering when determining how they allocate resources within their movement. Here it must be stated that the mistreatment and subjugation of nonhuman animals deserve to be included in relevant conversations about the environment even if some choose to ignore it. Lastly, Juneko Robinson employs some of Heidegger's ideas in her examination of the ways that clothing, and more specifically, "fast fashion," negatively impact the world we live in. For example, cotton cultivation drains water supplies and the monocultural production depletes the soil of nutrients.

Next, the politics segment of the book contains several chapters that explore a variety of other (sub-)issues. Jennifer Epley Sanders investigates the environmental attitudes of Muslims living in different countries and the intersection of their faith and thoughts on climate change as expressed in different policy statements. Suzanne Roberts analyzes the relationship that women have with nature as depicted in pieces of literature where the authors define the wilderness as a male space that is dangerous for women, and therefore, somewhere women do not belong. Camila Pombo studies the efforts that environmental nonprofit groups are making to engage communities of color about food sustainability. Her analysis of different programs yields a finding that some take a more "holistic" approach than others. Finally, John Duerk explores the positions of animal liberation and environmental activists who testified before Congress about the radical tactics their movements employed, and the federal government's response to them.

The last portion of the book consists of chapters that address matters of environmental policy—both policies adopted by governments and by private corporations. Markie McBrayer is interested in the proximity of bike lanes to neighborhoods of color and the relationship between bike lanes and the gentrification that has occurred in several American cities. In his examination of how some companies incorporate environmental awareness into their business models, Scott Lukas analyzes multiple themed spaces that have been designed to appeal to modern consumer interests. Elizabeth Koebele, Loretta Singletary, Shelby Hockaday, and Keri Jean Ormerod study two important river basins where water markets offset the effects of climate change to meet local needs. These cases studies have implications for similarly situated basins in other localities. Emilia Barreto Carvalho surveys the different regulatory approaches that governments may utilize to secure compliance from industry. Then, she measures the relationship between the "stringency" of the policies that countries adopt and the amount of greenhouse gasses (GHG) that they emit into the atmosphere. To conclude this anthology, Joe McBride explores how landscape paintings, drawings, and photographs played a role in generating awareness about the wilderness in the United States, and how that increased awareness informed and influenced efforts to protect such areas.

May the work contained herein yield greater understanding of the respective (sub-)issues and fuel some important conversations as they are intended.

–John A. Duerk, August 2020

Editor's note: The citations and references that appear in this anthology have been written in the seventh edition of APA style.

REFERENCES

Brenan, M. (2020, April 20). Updated environment data for Earth Day 50th anniversary. *Gallup*. https://news.gallup.com/poll/308552/updated-environment-data-earth-day-50th-anniversary.aspx

Funk, C. & Kennedy, B. (2020, April 21). How Americans see climate change and the environment in 7 charts. *Pew Research Center*. https://pewrsr.ch/2UqQsOI

Saad, L. (2020, April 20). Environmental ratings, global warming concern, flat in 2020. *Gallup*. https://news.gallup.com/poll/308876/environmental-ratings-global-warming-concern-flat-2020.aspx

U.S. Environmental Protection Agency. (n.d.). Plastics: Material-specific data. https://www.epa.gov/facts-and-figures-about-materials-waste-and-recycling/plastics-material-specific-data

U.S. Environmental Protection Agency. (2019, November). Advancing sustainable materials management: 2017 fact sheet. https://www.epa.gov/sites/production/files/2019-11/documents/2017_facts_and_figures_fact_sheet_final.pdf

Wilson, E. O. (1998). *Consilience: The unity of knowledge*. Vintage Books.

Part I

PHILOSOPHY

Chapter 1

Consumption and Consciousness

The Land Ethic Revisited

Mark Thorsby

Aldo Leopold has become a major figure in conservationist literature and now occupies a canonical position in most anthologies on environmental ethics. Published in 1949, "The Land Ethic" is the prize essay of A Sands County Almanac in which the author lays out a vision of ethics as being in a state of evolution. Like the biology of organisms and ecologies, the sphere of recognized moral obligations is evolving to include the natural ecological system that encompasses the systems of life overall. Despite its broad impact and reception by conservationists, professional philosophers have tended to be critical for a number of metaethical and practical reasons, some of which we will discuss in the next section. Overall, the reception of his work has tended to take one of the two tracks: on the one hand, critics have opposed "The Land Ethic" as bad philosophy, or on the other hand, philosophers have attempted to augment Leopold's vision of ethics with more robust philosophical support and arguments. This essay falls squarely in the later track by attempting to bring some of the insights from phenomenological philosophy to bear in our reading of the "The Land Ethic."

The philosophical reception to "The Land Ethic" over the past seventy years is but a small issue when we look to see whether or not Leopold's pronouncement that a new ethical recognition of the "land as community" would "penetrate our intellectual life" has actually occurred (1949). In other words, is anything like "The Land Ethic" actually evolving in the consciousness of modern people? At first blush, there is good reason to suppose so. Since 1949, the status of the environment as a moral concern has steadily gained traction and widespread acceptance, and yet, when we look at the status and health of the biosphere, the numbers seem to tell us otherwise. According to the United Nations IPBES Global Assessment Report on Biodiversity and Ecosystems (2019), the biodiversity and ecosystem functions across the globe are "deteriorating" and deteriorating fast. The authors of that study, composed of many leading scientists from around the world, write that the "rate of global change in nature during the past 50 years is unprecedented in human history" and these problems are accelerating (IPBES, 2019). As the human population has grown, more habitats are being lost, the ocean and air are getting more polluted, and the layers of biodiversity that line this planet are becoming thinner and thinner. The IPBES indicates that as the gross domestic product (GDP) of developed and developing nations rise, the extraction of living biomass for the purposes of consumption has steadily increased across the world. Consequently, "The global rate of species extinction is already at least tens to hundreds of times higher than the average rate over the past 10 million years and is accelerating" (IPBES, 2019). These kinds of statistics are so large that most of us are hardly able to comprehend the magnitude of the environmental decline across the planet's biosphere. The proof is in the pudding, as they say. Either Leopold was wrong and there is no new Land Ethic emerging in the conscience of modern peoples despite sustained lip-service to environmentalism or it is happening all too slowly. As E. O. Wilson put it in his 2002 book *The Future of Life*, nearly half of all species that existed at the beginning of the twenty-first century will likely be extinct by 2050. The clock is ticking, and it looks like "The Land Ethic" may be too late to the game. Can we make sense of this?

Leopold was fully aware that the impediments and obstacles to a Land Ethic were numerous and deeply rooted; they concern our attitudes and emotions, the education we have, and the economics of decision making. He writes, "[Perhaps] the most serious obstacle impeding the evolution of Land Ethic is the fact that our educational and economic system is headed away from, rather than toward, an intense consciousness of the land" (Leopold, 1949). His warning that we must work to disentangle our economic interests from the ecological outlook appears prophetically spot-on. But yet, Leopold's anticipation of the challenges facing a Land Ethic appears quixotically as stubborn as ever. He argues that we need a change in our community instinct that includes the land and the ecological systems of life. He writes that we need to stop "thinking about decent land use as solely an economic problem" (Leopold, 1949). Leopold's essay leaves out the mechanics of how a new way of thinking and seeing our obligations to the natural ecology might arise. The challenge is not simply our attitudes and instincts, but importantly, the perception of the land itself. If one can only perceive the land as an exterior resource in the first place, then clearly no deep abiding sense of the Land Ethic can take hold.

The goal of this chapter will be to highlight a way forward and break the logger-jam by suggesting some phenomenological paths forward. In particular, I argue that Leopold's "Land Ethic" requires a deeper analysis of the consciousness that accompanies our average everyday attitudes toward nature. In particular, a phenomenology of consumption can provide a more robust model for understanding the delay of the Land Ethic's emergence than merely pointing to egoistic financial interests. The economic attitude is a form of consciousness in which the relationship between a thinking person (and community) becomes singularly reduced in its exclusion of ecological relations. The economic attitude exemplifies a logic of instrumentalization. The type of attitude Leopold's Land Ethic envisions is one that maintains an intentional attitude toward the horizon of ecological relations and histories of dependency between ourselves, the land, and all things we call natural. Phenomenology, as a philosophy concerned with the meaning of our perception, can give us a better account of the problem, partially because it is one which can help to further ground the coherency of Leopold's conception of ethics as a community concept. As long as the instrumental logic of consumption pervades our capacity to perceive the land as otherwise, no Land Ethic will be forthcoming. We can potentially break the hold that consumption has on our perception of the natural world by emphasizing the history of the ecological world, and thereby bringing to life the community relations that are central to the Land Ethic. In what follows, I begin by reviewing Leopold's essay and the potential criticisms it sparks, then following the rhythm of the original essay we will explore the notion of ethics as an evolutionary phenomena, and the concept of consciousness and the economic attitude which bears on Leopold's community concept. Next, we briefly discuss the new agrarian movement and the implication that a poverty of the land ethics entails a poverty of land communities. Finally, we conclude with a phenomenology of consumption and the potential ramifications it might have for a land ethos.

THE ETHICAL SEQUENCE

It seems appropriate to begin with a short review of "The Land Ethic" before touching on its various criticisms. Let us begin by noting the overall thesis of his work, the structure of the essay, and his primary conception of ethics is in general; after which, we can turn to a brief discussion of the metaethical problems critics have found. First, Leopold begins by situating ethics as a process in development, an unfinished sequence that is widening over time to include more diverse forms of obligation than once imagined. Leopold opens "The Land Ethic" by recounting a scene from Homer's *Odyssey* in which Odysseus returns home from the Trojan war (after a very long and arduous journey), and the first thing he does is hang a dozen-slave girls on one rope that "he suspected of misbehavior during his absence" (1949). In other words, the epic hero Homer fashioned to demonstrate the ideal of Greek cleverness and action, Odysseus, has the character of a mass murderer and slaver. Readers of fine classical literature have a tendency to skip over these details in favor of more palatable literary fare. But Leopold jars us into the realization that ethics seems to have meant quite different things in the past than it does today. Although Leopold used Homer as his foil, we have numerous other examples to demonstrate the point. Romans practiced infanticide, the medieval Christian church believed that burning people alive would free and save souls, and the U.S. Constitution inscribed human slavery as a just and legal practice. In other words, the history of ethics is a pluralistic mess. Whether one agrees with the implication that ethical obligations are culturally relative, Leopold's point is a matter of fact. The ethical and moral obligations we take as obvious today (for instance, consider the United Nations Declaration on Human Rights) are the result of a historical process which, over time, has been extending and widening to include more and more of those who were previously excluded.

Leopold describes all of this as an ethical sequence. In particular, he likens the process to an evolutionary process. He writes:

> The extension of ethics, so far studied only by philosophers, is actually a process of ecological evolution. Its sequences may be described in ecological as well as in philosophical terms. An ethic, ecologically, is a limitation on freedom of action in the struggle for existence. An ethic, philosophically, is a differentiation of social from anti-social conduct. These are two definitions of one thing. (Leopold, 1949)

Leopold argues that ethics is the result of an evolutionary process. At its most basic conception, ethics refers to the ways in which people curtail their freedoms of action in order to achieve more noble ends in how they exist. This means that ethics offers us descriptions for both good and bad behaviors. In parallel fashion, the evolution of the ecology also differentiates good and bad behavior, but on a criterion of evolutionary fitness.[1] As ecological conditions change over time, the standards of evolutionary fitness similarly evolve. In this regard, the history of ethics can also be seen as an evolving process of differentiating good and bad forms of action and assigning behavioral prescriptions along the way. This may not be a complete picture of ethical theory, but it is a compelling one.

The description of ethics that Leopold gives is sociological rather than philosophical. Philosophers recognize the difference between social and antisocial behavior, but they are not concerned with simply advocating for some set of cultural norms, nor should an ethics be conceived of as simply an anthropological description. Perhaps the difference here is between having an ethic and being ethical. Additionally, we can distinguish between first- and second-order discourses in ethics. The first concern is practical, while the latter concern is conceptual or theoretical. By contrast, philosophers tend to focus on whether a practice is justified. Leopold does not offer a sufficient view of philosophical ethics, but this deficiency does not mean we should reject his point. Many philosophers, Hegel and Marx, for instance, argue that moral norms and social interaction are organized by a historical dialectical process.[2] I am not suggesting that Leopold endorses either philosopher, as far as I know he does not, but we should take this as evidence that the notion of an ethical sequence, or a dialectic of moral norms, is not an incoherent idea in itself. The linkage between evolutionary action and ethical value appears, to this humble thinker, as either an equivocation or too vague to fully make sense of. I think we can recognize parallels between both, but whereas ecological evolution is understood in purely descriptive terms, the values of an ethic are prescriptive in nature.[3] The point cannot be glossed over. The method of the ecologist is to recognize the implements of evolutionary fitness through a rigorous process of description and observation. This is only understood after the fact. By contrast, the nature of philosophical ethics is not simply to describe what it is that people value, but also, and perhaps most importantly, to identify the moral values that one ought to hold. For instance, imagine a society that holds and maintains an unjust practice within its society. For example, in the popular young adult novel *Hunger Games*, we are told the story of a postapocalyptic society that holds a festival each year in which innocent young men and women are forced to fight to death in a public spectacle. In this regard, the mere distinction between social and antisocial behavior becomes problematic. Clearly, the killing of innocent people is morally wrong, and yet, on the grounds that Leopold offers, social versus antisocial behavior, we are left with few resources by which to criticize the unjust practices of the unjust society. We might even imagine that this kind of activity increased the evolutionary fitness of the surviving community. But perhaps it is the cleavage between the description of how things are and the next evolutionary stage in moral development which supports Leopold's view.

The vision of the Land Ethic is one where the moral development of the human race is in a state of evolution and that as this process unfolds (which he says is inevitable) then the sphere of our moral obligations also enlarges so as to include the biotic community as a whole. Leopold defines the biotic community as the network of relations that living beings have with each other in a local ecological system. The farmer has a relationship with his crops, which in turn have a relationship with the insects and birds of a region, all of which depend upon the microorganisms that enable the fertility of soil. The mish mash and web of relations is best described as a network of interdependencies. Leopold's Land Ethic is a recognition of one's place and interdependency within the larger system. In ethical terms, the system is a community of which one is a citizen and member. The change toward the Land Ethic requires a change in the general consciousness of both the society and individual. This does not mean that there is no moral imperative that can be given to summarize this change in ethics. In fact, he offers the formula that a "thing is right when it tends to preserve the integrity, stability, and beauty of the biotic community. It is wrong when it tends otherwise" (Leopold, 1949). Callicott (1991), with a more robust history of ecological findings at his service, reframes this imperative by a similar formulation: "A thing is right when it tends to disturb the biotic community only at normal spatial and temporal scales. It is wrong when it tends otherwise." Yet, no matter the second-order moral formulation we might use to describe the principle at the heart of the Land Ethic, the overall vision is a historical dialectical view. I suggest that this is why Leopold struggles in "The Land Ethic" with the fundamental problem inherent to all historical dialectic narratives: how can new forms of social and ethical consciousness emerge when consciousness is dictated by historical conditions? There is a "chicken and egg" problem here. On the one hand, the emergence of a new set of ethical values can seem like a necessary next step in moral development; but on the other hand, since the consciousness of these values depends upon their emergence, it becomes too difficult to make sense of how they can emerge in the first place. Large sections of "The Land Ethic" are devoted to the problem of the problem of "economic" consciousness and the apparent stubbornness of people to see the land as simply an economic resource. Leopold discusses the Soil Conservation District Law of 1937, a law meant to supply farmers with the resources necessary to address the effects of the Dust Bowl and the loss of Wisconsin topsoil, if only they would participate. The law empowered farmers to write their own rules for land use, to be collectively enforced, and provided them with the necessary financial resources to enact their plans. Yet, no laws were written and the land continued degrading. Leopold laments the puzzling nature of the problem by consistently framing the problem in terms of economics and education. Why would a group of farmers who know the outcome that their practices, that they would destroy their own land, cut their own noses off by refusing to act? It seems that the immediate economic interests of individual farmers neutered their capacity to work in concert to protect the health of the land, the individual interest of each spoiled the community's capacity to act and no new rules were ever established. So despite the inevitable nature in which Leopold suggests the emergence of the Land Ethic, he nevertheless admits that "[land-use] ethics are still governed wholly by economic self-interest, just as social ethics were a century ago" and that each farmer is only able to act for his/her own personal immediate financial gain (1949). This is clearly related to the ethical problem of the commons, what Garrett Hardin (1968) calls, the tragedy of the commons. When a commons is used according to individual interests alone, then the commons is inexorably locked into a tragic logic of diminishment. Every individual agent will act according to its own interests only and thereby cause collective ruin for all. But the problem of an economic consciousness is not deep enough. To simply point the finger at economic rationality doesn't fully link up to the notion of an ethic in the fuller sense that I take Leopold (1949) to be suggesting when he says that "something as important as an ethic is never written," but (presumably) lived out in the social sphere through and through. Since the conditions upon which good and bad actions depend are consistently changing, an ethic that is simply written could never fully satisfy. Hence, since an ethic must first and foremost be lived, then it also indicates a specific attitude of perception. Some things are perceived as duties, others as infractions of right behavior, and still other behaviors are perceived as being morally neutral. Phenomenology, I think, can help deepen the story.

CONSCIENCE AS A PROBLEM OF CONSCIOUSNESS

The Land Ethic encourages a paradigm shift in the way in which we organize and recognize the various moral obligations we have, both toward other persons, but importantly toward the entire ecological system upon which we depend and are apart. Leopold calls this the community concept of the Land Ethic. Within the history of western ethical theory, the primary unit of obligation has been primarily placed with the individual. In Aristotle's virtue ethics, the happiness of the individual is what guides one toward virtuous conduct. In Kant's deontological philosophy, only individuals have moral value, and inherent moral worth. In the utilitarianism of John Stuart Mill, right up to that of Peter Singer, the individual's preferences, even when considered against the majority, still maintain the logical priority of analysis. One cannot determine the greatest amount of happiness for the greatest amount of people, if they do not first have a requisite understanding of the individual's happiness. In fact, the priority of the individual within moral theory is so pervasive and widespread, that it really is the dominate position in moral philosophy. Even moral subjectivists—those who think that moral values are merely subjective mental experiences (and/or disposition)—side with the priority of the individual. We might say that the individual, as the unit of analysis, reveals what might be called the individualistic prejudice of western moral theory. This is where, at least in my reading of Leopold, that we can locate a radical departure from most moral philosophers. Two readings of Leopold are possible; either a conventional view or the radical view in which the Land

Ethic represents an overturning of former systems of obligation. This may be contributing to the underlying reason why so many philosophers have criticized Leopold. Environmental ethicists describe this orientation in terms of holism. That is, the primary unit of moral consideration does not refer to the individual, but to the whole matrix of relations. For Leopold, the community as an aggregate whole is the primary unit of moral analysis. The individual is a biotic citizen who holds obligations derived from the whole—not the reverse. Hence, the problem of individual self-interest and the possibility of the Land Ethic really emerging within the minds and hearts of her biotic citizens requires a conceptual movement from the individual to the whole. And this ethic, or ethos, must have a different logic in consciousness from that of the singular individual's interests. Phenomenological philosophy may provide some means to clarify the problem.

Phenomenology is the study of consciousness and the necessary structures in consciousness such that it is possible to be conscious at all. Unlike psychology, phenomenology seeks the articulation of the underlying conditions for consciousness in an attempt to make sense of how meaning arises; the principles for consciousness can be deductively derived from experience. Phenomenology, as a discipline, was founded by the German philosopher Edmund Husserl (1859–1938). Although he is little known outside of the circles of philosophy departments, Husserl has had a profound impact on the course and development of the twentieth-century philosophy, providing the impetus for the existentialist philosophies of Martin Heidegger, Jean-Paul Sartre, Hans-Georg Gadamer, and Jacques Derrida among many others. Phenomenology though is not a theory about consciousness but rather a method for describing states of awareness. Phenomenology begins with the insight that human consciousness is always intentional.

Intentionality is the directedness of consciousness, and this property is always found in consciousness no matter the mental state or object of one's awareness. Consciousness ie never conscious of nothing, or everything, but always focused on a particular phenomenon in consciousness. While you read this chapter, your intentionality is directed toward the meaning of these words written on the page, but as soon as you hear a noise, the direction of consciousness shifts. At no point does intentionality disappear; for if it did, then one would be unconscious. Using this property as a methodical guide, Husserl would develop a method of analysis he referred to as phenomenological reduction. In order to examine the role of perception itself, Husserl would begin with a phenomenon and then exclude (or bracket out) the assumption that the phenomenon had an actuality outside of consciousness; so as to then describe its features in consciousness but as an object of consciousness. He might suspend his assumption that the chair, for instance, has a reality, so that he could then describe the phenomenon as a phenomenon; rather than as an object in the world. This was the essence of his phenomenological method; but as a method, it works by reducing our forms of intentional awareness to their essential elements. If intentionality is the directedness of consciousness, then every directedness always includes an exclusion. When the farmer is directing his consciousness toward a crop, he simultaneously excludes all other objects toward which he could be aware of (e.g., his horses). This occurs even within our field of vision with immanent phenomena. One is never fully aware of everything in their line of sight, but rather, one becomes aware of things either one by one or as combined generalities. This feature of our intentionality allows one the possibility to exclude, or parenthesize, aspects of our awareness, so that we can better understand the structure in which our awareness depends. In fact, it is this possibility which allows one to have different sorts of attitudes. This is particularly important for the consideration of the economic attitude with which Leopold is addressing. As a mathematician, Husserl (1983) considered what he called the arithmetical attitude, whereby one can exclude features of their experience to focus solely on the pure ideas of mathematics. We can, and should, suppose that there is also an economic attitude. Philosophically, Husserl was interested in understanding the meaning of the ideas in mathematics, how they can be made conscious, and in what sense they might exist. He argued that when we take on differing attitudes, we appropriate to ourselves differing "worlds" of awareness. Each world of awareness has its horizon of possibilities for cognizing the phenomena, its past, its present, and its future. The mathematician is in a world of numbers, while the economic attitude appropriates an economic world of instrumental exchange. Yet, beneath all of these attitudes is a natural attitude. He writes:

> The arithmetical world is there for me only if, and as long as, I am in the arithmetical attitude. The natural world, however, the world in the usual sense of the word is, and has been, there for me continuously as long as I go on living naturally. As long as this is the case, I am "in the natural attitude," indeed both signify precisely the same thing. That need not be altered in any respect whatever if, at the same time, I appropriate to myself the arithmetical world and other similar "worlds" by effecting the suitable attitudes. In that case the natural world remains "on hand": afterwards, as well as before, I am in the natural attitude, undisturbed in it by the new attitude. (Husserl, 1983)

We can think of the various attitudes we can take in our awareness, both of ourselves and things around us, as layers of intentional exclusion mapped onto the natural world. These layers organize both interests and perceptions by structuring our relationship to phenomena. We can inhabit the world of philosophy, mathematics, sports, conservationism, or religious life, but always on the condition that there is a natural world to begin with. Husserl is, of course, speaking analogically of experiential "worlds." This idea can really go a long way in helping us make sense of the economic attitude that seems to plague the promise of the Land Ethic. Since the worlds of our intentional experience are built upon a natural attitude, that in turn depends upon our "living," and thus that there be a natural world in which to live, then the economic attitude coexists with the perceptions we have of nature. It is a puzzling problem. The land developer and environmental activist both see the same land, but with differing orders of meaning. It is as if our economic attitude supervenes itself and colors the consciousness we might have toward nature. This is, I think, a phenomenological description of the problem facing the emergence of the Land Ethic.

Central to Leopold's own discussion is the problem of education. It appears that more environmental education (i.e., more ecological knowledge) is at best insufficient to motivate action or prompt a change of attitude; at worst, the call for more education (or ecological research) can serve as an excuse to avoid change altogether. Yet, the phenomenological picture can provide a more insightful view. Education about nature, at least by itself, is insufficient to change an attitude that inherently acts by excluding nature so as to render a wholly economic world. This also helps explain how the A-B cleavage of conservation is possible. Similar to the Deep Ecology Movement of Arne Naess, Leopold distinguishes two types of conservation. On the one hand there are (A) those conservationists who are merely concerned with the forest as a resource to be maintained and there are (B) those who concern themselves with the conservation and preservation of natural biotic systems for its own sake. In other words, there are conservationists with an economic attitude, and there are those with the Land Ethic. I suppose the Land Ethic is its own type of attitude in consciousness as well. It is an attitude that takes the form of an ethos or habitual pattern of action. The Land Ethic also rests upon our primary form of being in nature, simply being in the world, enjoying the spatio-temporal field of beings that lie before us. The Land Ethic is a development of consciousness that excludes the economic exclusion of nature, so as to enable a perception of our place within a living biotic community.

INTERSUBJECTIVITY AND THE COMMUNITY CONCEPT

One of the central ingredients of the Land Ethic is the community concept. He argues that at the core of the ethos is the importance of the community. Of course, Leopold does not mean community in an anthropocentric manner. His "community" refers to the biotic, or ecological, community that sustains life itself over time. Leopold offers us the image of a land pyramid. He writes,

> An ethic to supplement and guide the economic relation to land presupposes the existence of some mental image of land as a biotic mechanism. We can be ethical only in relation to something we can see, feel, understand, love, or otherwise have faith in . . . A much truer image is the one employed in ecology: the biotic pyramid . . . Plants absorb energy from the sun. This energy flows through a circuit called the biota, which may be represented by a pyramid consisting of layers. The bottom layer is the soil. A plant layer rests on the soil, an insect layer on the plants, and so on up through various animal groups to the apex layer. (Leopold, 1949)

Between the various layers are lines of dependency and relation in a complex web of relations that occur over time and at different time scales. Leopold offers the principle of balance between these layers as being a source of value. Without the complex web of interdependencies, life would not be possible. Thus, each axis of relation forms a node of life, through which energy circulates through the biota. The rhythm of Leopold's argument is to highlight that a Land Ethic emerges when one begins to recognize themselves within those community of relations. As Callicott highlights, the time scale of a relation is important for understanding the proper sorts of action that promote or diminish the land biota. Ecological relations within the biotic web have their own histories and future effects. The point is not an altogether foreign idea to the phenomenologist.

One of the central tenets of Husserl's phenomenology was that even though our modes of consciousness can appear isolated, solipsistic, and individualistic, the various attitudes in consciousness are always conditioned a priori by the community of which we are a part. Husserl calls this condition intersubjectivity. In the Cartesian Meditations he offers a very complicated account of intersubjectivity as a community relation between monads. A monad refers to other beings with subjective experience. He writes that the "constitution of the world essentially involves a harmony of the monads" (Husserl, 1960). For instance, our awareness of objectivity is based on the intersubjectivity we have with others in our community. That is, our ability to recognize something as objective depends upon the form of agreement we have with other persons, other subjects, in our judgments about the world. Intersubjectivity refers to the relations that exist between subjects. So, for instance, the ecologist will not consider their scientific findings objective, unless their work can be reproduced by others. The only way in which we can know that a phenomenon is objective is when it is also shared by others. This fact grounds the possibility of the meaningful assertions we have with others. A child cannot learn language by themselves. The conditions for our experience and awareness of the world are intersubjectively based, and these conditions form the basis upon which our culture and ethics can emerge. Many of these conditions emerge slowly over time within the traditions of a culture, and may not always be present to consciousness but sedimented into our baseline habits and assumptions about the world. Yet, all of these intersubjective conditions are fortuitously founded upon an intercourse with nature. He writes, “Everyone, as a matter of apriori necessity, lives in the same Nature, a Nature moreover that, with the necessary communalization of his life and the lives of others, he has fashioned into a cultural world in his individual and communalized living and doing” (Husserl, 1960). Thus, the attitudes we bring to bear on nature are culturally based on a shared natural world. Like Leopold, Husserl argues that we are members of nature. He writes, “My animate organism is the central member of ‘Nature,’ the ‘world’ that becomes consisted by means of governance of my organism” (Husserl, 1960). In this sense, nature is the foundational phenomenological world of which we are apart. Leopold calls us to join the biotic community as citizens. It is striking that even Husserl uses a metaphor of the political to describe this relationship. Although Husserl tends to maintain a strict distinction between the world of nature and the cultural world, there are strong resonances between the two thinkers, such that a phenomenology of the Land Ethic is consistent to the insights of both thinkers. Even when we return to the “cultural world, we find that it too, as a world of cultures, is given orientedly on the underlying basis of the Nature common to all” (Husserl, 1960). Based on this, we can widen the concept of intersubjectivity to include the world of nature too, upon which intersubjectivity depends. To some extent, our language fails us here since nature need not include subjects. But the foundation for ethics will always be intersubjective, and since the intersubjective is dependent upon the natural, then what the Land Ethic does is to prompt our intersubjective communities to recognize their community as being in intercourse with nature itself. In order to recognize the biotic community we have be become acquainted with the various time scales at which the intersubjective relations between human beings and nature unfold. Thus, ethics can be made to incorporate that upon which it depends. The Land Ethic is “the existence of an ecological conscience” and “a conviction of individual responsibility for the health of the land,” and is this not a way of describing a community that evolves to see itself as member of an even larger community of nature? (Leopold, 1949).

CONSCIOUS CONSUMPTION

Today, both cultural communities and the communities of nature have been diminished by the economical attitude. Wendell Berry, a well-known agrarian poet and author, has written extensively on both. In *The Art of Loading Brush*, Berry offers a vision of the new agrarian, those concerned with the protection and revitalization of rural communities that depend upon the land; both of which have been victims of economically minded decision making. Berry (2017) reminds us, "We have, in fact, been turning our country into an economy as fast as possible, and we have been doing so by an unaccounted squandering of its actual, its natural and its cultural, wealth." He sees this both as a human and natural catastrophe. At the center of the new agrarian movement is the recognition that we, as human beings, have become displaced from nature and that this displacement has curtailed a recognition of its value. The new agrarian seeks to restore the possibility and dignity of a life on the land, and in relation to the land. His writings are highly influenced by Aldo Leopold whose

> conception of humanity's relation to the natural world was eminently practical, and this must have come from his experience as a hunter and fisherman, his study of game management, and his and his family's restoration of their once-exhausted Sand County farm. He knew that land-destruction is easy, for it requires only ignorance and violence. But the obligation to restore the land and conserve it requires humanity in its highest, completest sense. (Berry, 2017)

Like Leopold, Berry points in the direction of an economic mentality, and the ferocious appetite for consuming nature, as being central to the problem. Unlike Leopold, Berry sees the agrarian project as a form of restoration, a way of dialing back the clock on the economic consciousness that pervades the contemporary consciousness. He writes, "The tradition of cooperation with Nature has persisted for many centuries, through many changes, sustained by right principles, the proofs of experience and eventually of science, and good sense. Its diminishment in our age has been tragic. If ever it should be lost entirely, that would be a greater 'environmental disaster' than global warming" (Berry, 2017). Either way, the problem at the root of the rot is the metastasis of an economic attitude that is a simplistic instrumental relation to the phenomena of experience. The project of the new agrarian depends on an overturning of consciousness.

The economic attitude is first and foremost an attitude of instrumentalization. Take the experience of buying produce at a local grocery market. There are many kinds of shoppers, some who clip coupons, others who spend as much as they can, but if we were to think of the shopper who has a thoroughly economic attitude, then we would be imagining a shopper who buys as much as possible for the least amount possible, as efficiently as possible, at any immediate moment. But the economic attitude would not simply be one that seeks a good deal, but rather that attitude which values the produce according to its exchange value. The produce purchased has its value as a "good buy" rather than as a "nutritious meal." This attitude reduces the relation we have with products to a single relation of exchange value, characterized by a relation of consumption. The role of the produce in its natural environment is entirely invisible for the purveyor of the economic attitude. Thus, the economic attitude takes the objects of the market as merely consumables to be valued on their financial instrumentalization. We mentioned above that every attitude is like a layer of exclusion that colors our perception of the natural objects we encounter. This means that the economic attitude excludes other sources of value beyond the monetary. The temporal relation is reduced to the immediate moment. Where that produce has come from or what effect its production has had on the ecological landscape is never asked. The farmers who produced the fruit are completely excluded in the value of the produce. The pesticides, and the intractable damage they cause the land, are also excluded from consideration. Finally, the effects of one's purchase are also excluded from consideration so long as the price is right. Consider the amazing switch in consciousness that is possible when we enter a grocery store. All the products are packaged nicely in a fine layer of plastic, entombed within their plastic and cardboard casings. All of this appears perfectly normal and clean to the consumer. Yet, after the consumer buys their products, they simply strip the packaging off and discard it as trash. What a switch in consciousness! The very same item that was considered clean and acceptable in the marketplace becomes dirty trash to be simply dispensed with. Imagine if everyone walking through a large grocery store or supermarket came to the realization that the isles were lined with nothing but trash, head to toe. In this sense, the economic attitude powerfully renders our conscious awareness of the world into immediate sterilized objects of use. When it comes down to it all, the economic attitude treats all natural objects as commodities to be consumed via a principle of instrumentalization; that which comes before the consumption and that which comes after is excluded from the horizon of our experience. With this attitude, the land is not seen as an ecological fountain of energy distribution, but rather as something to be developed. What came prior, the history of the land complete with its natural ecological niches and stories is reduced to a field for clearing. What comes later, the utter destruction of the habitats and the incredible loss of species, is reduced to an economic externality barely noticed. The Land Ethic requires a paradigm shift in attitude, and an awareness of the temporal relation we have with the land.

That consciousness which acts as a barrier to the Land Ethic is the consciousness of consumption. A biotic citizen cannot see her only relation to the biotic community as one of consumption, but rather in terms of a prolonged interdependency. Thus, the Land Ethic's emergence requires an overturning of the perception that all natural things exist for consumption. This is an important point for contemporary environmental ethics and for sound environmental policy decision making. When faced with environmental crises, a popular strategy is to turn toward technological solutions. Instead of ending coal power production, we seek its sequestration. Instead of letting land lie fallow, we stock it with artificial fertilizers. Instead of eating less meat, we genetically modify whole species to get fatter quicker. Indeed, no technological solution can feed the beast of consumerism enough to protect the land. A true solution to our ecological catastrophe requires simply that we simply stop consuming it. Yet, for many modern persons, this is simply unimaginable. Today, a modern standard of living has become synonymous with the unsustainable depletion and consumption of the natural world. "In the past 50 years, the human population has doubled, the global economy has grown nearly fourfold and global trade has grown tenfold, together driving up the demand for energy and materials" (IPBES, 2019). E. O. Wilson (2002) estimates that if everyone on Earth were to enjoy the level of consumption that modern Americans take for granted more planets are needed. There is quite literally not enough of the Earth to feed our economic and consumptive attitudes if the present trends continue. Leopold's Land Ethic requires a renewal of attitude and a change away from the intentionality of consumption.

THE ETHOS OF THE LAND ETHIC

A Land Ethic, like any lived ethic, not only helps us understand what we should do, but it also helps us actually do it. Ludwig Wittgenstein once wrote that if there should ever be a book written about ethics that it would explode, for ethics taken in an existential register requires that we take the course of actual living seriously. A shift away from economic consciousness requires a shift of consciousness, and the reduction of all objects we encounter (both those in nature and in our cultural communities) become instruments. It is one thing to consume something in the world, it is entirely another to see the world as something to be consumed. Yet, a mere shift in consciousness is moot if it cannot render actual change. The Land Ethic, or any environmental ethic, is a failure if the biotic community collapses, regardless of whether or not it is rational and sound philosophical theory. So, we cannot end the discussion of the Land Ethic here.

I am reminded of the problem Aristotle faces, in his *Nicomachean Ethics*, of how it is that one can become a just person. The problem is equally soluble here. How is it that one can enact a change in our own awareness if we are not already aware of it? And by awareness, I do not mean educational knowledge. Aristotle (1984) says quite simply, a just man becomes just by doing just acts. Perhaps the advice for individuals here is that we need to stop seeing the world as something to be used and consumed. We can mitigate the economic attitude we have toward nature by finding ways to disengage an attitude of instrumentalization. We noted earlier that economic rationality has a horizon which excludes the natural histories of our objects. One way, suggested indirectly by Nietzsche, is that we should get into the habit of reminding ourselves that all the objects within our experience have a history. This is true for both living and nonliving things. The land and ice also have a history, but so do the objects we buy. We must habituate an ethos, a habit of life, that recognizes the relations all beings play in nature; this occurs by seeing the relations we have with natural objects over time. If we can recognize that the natural objects in our perception—all of which are iterate forms of nature—have relations beyond the linear form of consumption, then I think we can begin to diminish our instrumentalization of these objects. Knowing the history of an object can neutralize, in consciousness, our tendency to see nature as merely an object to be consumed. For instance, consider a hog on display at a local county fair. For those looking to buy the hog, the perception of the relationship they have with the animals is purely instrumentalized according to the purposes of the buyer—making bacon, for instance. But as soon as one considers the history of the animal, that it too is the child of a mother, or that it too has had its own experiences of a life, then suddenly the singular relation of instrumentality gets disrupted. Readers of *Charlotte's Web* recognize Wilbur as a being of value and not simply an agricultural product precisely because Wilbur has a story—he has a history. I suggest that we should pursue a similar approach to nature and work toward cultivating the histories of natural objects. This might help to disrupt and suspend the economic attitude of consumption by reframing our intentionality of the relations that extend beyond our own uses toward an awareness of the intersubjective relations that exist in nature. Perhaps an awareness of these relations can enable an awareness of one's own place and citizenship within the wider field of the community of nature.

The Land Ethic concerns the type of relation we have with things. The economic attitude assumes a one-to-one relation in which my economic interest is central to the logic of considerations. By recognizing the history of natural objects, the relationship we have to natural objects shifts in consciousness around a different organizing principle. A history reinvests an object with a sense of value because the object becomes the center of its own story. For example, when one is confronted with the history of where their food comes from, when confronted with its history, then new values can emerge and the instrumental logic gets excluded. But the harder answer is that to avoid a worse ecological catastrophe we just need to just stop consuming the natural world. Every ethics requires its own set of habits and habitual conduct. We must habituate nonconsumption and a consciousness that recognizes itself within the larger biotic community. We must begin to value the things of nature beyond our individual economic interests. Our values can change, they have changed before. So how does the Land Ethic emerge? By acting ethically toward the land and by first perceiving nature as something existing for its own sake.

NOTES

1. Evolutionary fitness refers to the ability for the members of a species to procreate. Those members of a species which can procreate before death, and thereby continue the gene pool of a species, are considered evolutionarily fit. Members of a species which do not, or cannot, procreate, and evolutionarily unfit and get excluded from the species gene pool over time.
2. A historical dialectic refers to the baseline logic implicit within the events of a larger historical progression. Hegel thought of the dialectical process as a movement of negating the negated thesis.
3. This point is also made by J. Baird Callicott's essay "The Conceptual Foundations of the Land Ethic."

REFERENCES

Aristotle. (1984). Nicomachean ethics. In J. Barnes (Ed.) *The complete works of Aristotle*. (Vol. 2). Princeton University Press, 1729–1867.

Berry, W. (2017). *The art of loading brush: New agrarian writings*. Counterpoint.

Callicott, J. B. (1991). The land ethic. In D. Jameson (Ed.) *A companion to environmental philosophy* (pp. 204–217). Blackwell Publishing.

Callicott, J. B. (2010). The conceptual foundations of the land ethic. In C. Hanks (Ed.) *Technology and values: Essential readings* (pp. 438–453). Wiley-Blackwell.

Hardin, G. (1968, December). The tragedy of the commons. *Science, 162*(3859), 1243–1248.

Husserl, E. (1960). *Cartesian meditations*. Translated by Dorion Cairns. Martius Nijoff Publishers.

Husserl, E. (1983). *Ideas pertaining to a pure phenomenology and to a phenomenological philosophy first book*. Translated by F. Kersten. Martinus Nijoff Publishers.

IPBES. (2019). Summary for policymakers of the global assessment report on biodiversity and ecosystem services of the Intergovernmental Science-Policy Platform on Biodiversity and Ecosystem Services. https://doi.org/10.5281/zenodo.3553579.

Leopold, A. (1949). *A Sands County almanac and sketches here and there*. Oxford University Press.

Wilson, E. O. (2002). *The Future of Life*. Vintage Books.

Chapter 2

Rawls and the Distribution of Human Resources by Those in the Animal Rights Community

Alan C. Clune

PREPARATORY REMARKS ON THE TITLE AND ARTICLE

In the title, by "human resources" I mean activist work and monetary resources of those in the animal rights community (ARC). The ARC includes all those who seek to abolish all human behavior which uses animals[1] primarily for human benefit. Examples of the categories of animal use the ARC is concerned to abolish include the following: factory farming for animal products of any kind; researching on animals for any purpose; hunting animals for food, sport, or any commercial purpose; and using animals in entertainment.

The ARC can be seen as activists whose views are most in line with animal rights philosophers such as Tom Regan (1983), Mark Rowlands (2009), and Gary Francione (2008). Although there are disagreements among these philosophers, they do agree that animals have the prima facie right to be left alone comparable to the same right humans are taken to possess. They therefore agree that we ought to abolish all use of animals which causes them suffering and/or death primarily for the benefit of humans.

If these philosophers are right, then the next step is to devise a plan to bring about abolition. Hence, this article is concerned with determining the proper way to distribute ARC resources in furtherance of abolition. I make a case for a principled manner of distribution which is informed first and foremost by concerns about animals' rights.

Since the ARC is interested in a distribution of resources that is respectful of animals' rights, then the ARC ought also to be interested in a distribution that is fair to the animals of concern to them. For this reason, I will look to *A Theory of Justice* by John Rawls (1999) to argue for a just distribution of ARC resources. I am not interested in arguing that animals would be covered generally by Rawls's theory of justice as formulated by Rawls. I do not think they would be, even though I do think that animals would be covered under a broader interpretation of Rawls's theory, such as what is found in *Animal Rights: Moral Theory and Practice* by Mark Rowlands (2009).

There has been much discussion on whether Rawls's theory of justice can be extended to cover animals. The latter discussion deals primarily with the question of whether the reasoning used by Rawls which leads to justice for humans can be *extended* to grant coverage to animals *as well*. As stated above, I think that it can if more broadly interpreted. Answering this question, however, involves how to work out a Rawlsian theory of justice that covers generally a group which includes *both* humans and animals. This is distinct from the more modest question addressed herein of whether certain of Rawls's techniques for determining fairness can be applied to a group of beings which does not include humans at all. In this chapter, although I answer the latter question, my argument does not prima facie depend on, or seek to answer, the former question.

Hence, below I will be interested in employing Rawls's concept of the original position in the form of the veil of ignorance to determine a fair distribution of resources by the ARC. My strategy is to make use of the fact that Rawls's starting point for his theory of justice is a clearly defined society of concern, a society composed solely of humans. As we will see, essential features of his chosen society of concern play a fundamental role in guiding those behind the veil of ignorance to his principles of justice.

In a similar way our starting point will be our own society of concern, a society composed solely of animals. Our society of concern, as we will see, is different from Rawls's in its essential features. As with Rawls's derivation of principles of justice, the essential features of our society of concern play a fundamental role in guiding those behind the veil of ignorance to our principles of justice for the distribution of ARC resources.

This is in large part an argument by analogy to Rawls's manner of arriving at principles of justice. Once one has an adequate description of one's society of concern, and only then, can one reason to principles of justice for that society. We will find that since our societies possess different essential features, we will be led to different principles possessing largely different meanings. By way of fully drawing out the analogy, I will include an analysis of why the different sets of essential features lead to the different principles.

THE NATURE AND PURPOSE OF THE VEIL OF IGNORANCE

In *A Theory of Justice*, Rawls (1999) has us imagine an original position where rational agents, each acting in one's own interest, deliberate together in order to come to agreement on the rules for governing their society, including deciding on the rules for the distribution of resources within society.

In the interest of fairness, these agents are placed behind a veil of ignorance during their deliberations to determine society's rules (Rawls, 1999). Rawls employs the veil of ignorance to bracket out of consideration particular properties (such as intelligence, strength, and social status) of the individuals during their deliberation. The underlying reason for this is that these properties are undeserved. They are contingent features either of one's nature or of one's circumstances in society. Undeserved properties, according to Rawls (1999), should not in themselves place an individual in a place of advantage or disadvantage in society.

The idea here is that if a society is to have just fundamental rules, individuals ought to be treated as though they are morally equal for the purposes of coming up with these rules. To treat individuals as morally equal requires acknowledgment that circumstances beyond individuals' control ought not automatically to disadvantage or advantage them by being somehow constitutive of the rules devised for society. For example, the rules should not place individuals at an automatic disadvantage due to the color of their skin. This would be to penalize them for circumstances that are beyond their control. This would be unfair.

The veil of ignorance, then, ensures that no one uses one's knowledge of one's own underserved particular properties in order to sway the deliberation process disproportionately to one's own advantage. As Rawls (1999) states, "The original position [being behind the veil of ignorance] is the appropriate initial status quo which insures that the fundamental agreements reached in it are fair."

There is an abundance of information that Rawls (1999) allows access to, however, behind the veil of ignorance: "The only particular facts which the parties know is that their society is subject to the circumstances of justice and . . the parties are presumed to know whatever general facts affect the choice of the principles of justice." The circumstances of justice are, roughly, the conditions that make a theory of justice compelling. Knowledge of these circumstances and general facts do not place any individual at an unfair advantage or disadvantage in the negotiations because it is either too general or would apply to everyone equally.

HOW RAWLS'S SOCIETY'S ESSENTIAL FEATURES LEAD TO HIS PRINCIPLES OF JUSTICE

In this section, I explain how Rawls applies the veil of ignorance over his chosen society of concern. As mentioned previously, I will accentuate how certain essential features of Rawls's chosen society of concern frame the deliberation process behind the veil of ignorance, leading to his particular principles of justice.

Let us begin with a description of Rawls's society of concern. It is clear that Rawls's (1999) society of concern "is a cooperative venture for mutual advantage" "[among] free and rational persons concerned to further their own interests." The following three essential features of this society of concern are as follows:

Feature 1: The society consists of "free and rational persons concerned to further their own interests" (Rawls, 1999).

Feature 2: The society is a cooperative venture among its members which generates benefits which are greater than what individuals could achieve on their own (Rawls, 1999).[2]

Feature 3: Society's cooperative venture promises mutual benefit to its members.

It is clear from reading Rawls that these essential features are known from behind the veil of ignorance. This is because those behind the veil of ignorance know the circumstances of justice (Rawls, 1999). These circumstances of justice include that members of society will be rational persons with their own life plans (Rawls, 1999).[3] These circumstances also include that it will be in the interests of those in society to cooperate because it will produce mutual benefits that are otherwise unattainable (Rawls, 1999).

According to Rawls (1999), his society of concern has a basic structure which is the primary subject of justice: "For us, the primary subject of justice is the basic structure of society, or more exactly, the way in which the major social institutions distribute fundamental rights and duties and determine the division of advantages from social cooperation." Now, it is possible to have a basic structure that is unjust. Rawls's goal, then, is a basic structure that is designed to be in accord with his principles of justice. Toward this end Rawls (1999) employs the veil of ignorance over his society of concern, arriving at the following principles of justice:

1. "Each person is to have an equal right to the most extensive total system of equal basic liberties compatible with a similar system of liberty for all[,]" and

2. "Social and economic inequalities are to be arranged so that they are both (a) to the greatest benefit of the least advantaged,[4] consistent with the just savings principle,[5] and (b) attached to offices and positions open to all under conditions of fair equality of opportunity."

Principle 1 grants the following sorts of liberties to members of society: "Political liberty (the right to vote and to hold public office) and the freedom of speech and assembly; liberty of conscience and freedom of thought; freedom of person . . . the right to hold personal property and freedom from arbitrary arrest and seizure" (Rawls, 1999).

Principle 1 would be chosen from behind the veil of ignorance due to the first feature of Rawls's society of concern: that society consists of free and rational persons concerned to further their own interests. From behind the veil of ignorance one would understand that these sorts of liberties would be required in order that persons be able freely to act to further their own interests.

Furthermore, as Rawls (1999) points out, since individuals' "fundamental aims and interests are protected by the liberties covered by the first principle, they [that is, those behind the veil of ignorance, would] give this principle priority." Consider that individuals' "interests, including even their fundamental ones, are shaped and regulated by social institutions [that constitute the basic structure]" (Rawls, 1999). Therefore, individuals, from behind the veil of ignorance, would want to choose and give priority to the liberties granted by principle 1 in order that these individuals maintain control over the social institutions that will shape and regulate their ability to further their own interests (Rawls, 1999). Feature 1 implies the need for principle 1.

Principle 2b would be chosen from behind the veil of ignorance due to feature 2 of Rawls's society of concern: that society is a cooperative venture among its members which generates benefits which are greater than what individuals could achieve on their own. If society is to be a cooperative venture, then it is principle 2b that establishes society as a cooperative venture by providing the rules by which society's cooperative venture are to be formed.

There are two reasons—associated with feature 2—why principle 2b would be chosen from behind the veil of ignorance. First, individuals would understand, from behind the veil of ignorance, that cooperative ventures are capable of yielding greater benefits than when individuals work alone. So, a cooperative venture promises a better life than a noncooperative venture.

Second, individuals behind the veil of ignorance would understand that principle 2b helps avoid arbitrariness entering into the process by which individuals become employed in the cooperative scheme by ensuring that competition for positions occurs "under conditions of fair equality of opportunity." As Rawls (1999) points out, those who might be left out due to unequal opportunity "would be right in feeling unjustly treated . . . because they [would be] debarred from experiencing the realization of self which comes from a skillfully devoted exercise of social duties."

Now, by way of showing why principle 2a would be chosen from behind the veil of ignorance, we have to say a bit more about principle 2b. According to Rawls (1999), the role of the principle of fair equality of opportunity, and hence, the role of 2b, is to "insure that the system of cooperation is one of pure procedural justice." Pure procedural justice can be understood as a "correct or fair procedure such that the outcome is likewise correct or fair, whatever it is, provided the procedure has been properly followed" (Rawls, 1999).

Rawls illustrates this kind of justice using the case of gambling. As long as all the background conditions to the game are fair (the rules are clear, the bets are freely made, no one cheats, etc.), then the outcome is fair no matter what it turns out to be. Fair equality of opportunity—and therefore principle 2b—likewise sets up fair background conditions against which individuals achieve their places within the cooperative scheme of activities.

Now, according to Rawls (1999), one wants also to treat "the question of distributive shares as a matter of pure procedural justice." This is accomplished by principle 2b. In order to see this consider that what determines the correct distribution of shares is a just basic structure, one that is in accord with the principles of justice (Rawls, 1999). And according to Rawls (1999), it is principle 2b, among these principles of justice, which plays the essential role in distributive justice. Rawls (1999) states, in discussing the role of pure procedural justice in distributive shares, "Thus, in this kind of procedural justice the correctness of the distribution is founded on the justice of the scheme of cooperation from which it arises." Now, since principle 2b gives rise to a just scheme of cooperation, and as Rawls states, the justice of this scheme plays a foundational role in the correct distribution, it follows that principle 2b plays an essential role in a just distribution. It therefore also follows that individuals' shares in society's benefits will hinge on individuals' places within the cooperative scheme.

Now we are prepared to explain why principle 2a would be chosen from behind the veil of ignorance. As it turns out, individuals' fair share in society's benefits does not hinge only on their places in the cooperative scheme determined by principle 2b. Rawls (1999), referring to his preferred conception of principle 2 as the "democratic conception," points out that the "democratic conception holds that while pure procedural justice may be invoked to some extent at least, the way previous interpretations do this still leaves too much to social and natural contingency." What this means is that while principle 2b ensures a role for pure procedural justice in determining distributive shares, there is a need to constrain the influence of social and natural contingencies on the distribution. Principle 2a is what provides this constraint.

Principle 2a would be chosen from behind the veil of ignorance due to feature 3 of Rawls's society of concern: that society's cooperative venture promises *mutual* benefit to its members. Part 2a is Rawls's (1999) difference principle, and it is designed so that society "does not weight men's shares in the benefits and burdens of social cooperation according to their social fortune or their luck in the natural lottery." Hence, the difference principle acts as a limiting condition on part "b" by mitigating fortuitous circumstances which create better situated individuals. And this avoids principle 2b leading to a meritocratic way of distributing men's shares in the benefits from social cooperation, where these benefits of society accrue only to the most talented (Rawls, 1999).[6] A meritocratic method of distributing such benefits would "leave too much to social and natural contingency." Hence, principle 2a ensures that society's cooperative venture has the goal of mutual advantage rather than the goal of distributing benefits of social cooperation using a meritocratic method (Rawls, 1999).[7] One would choose principle 2a in order to avoid being left out of sharing in society's benefits due to natural or social contingencies that are beyond one's control.[8]

The above analysis has given us an explanation of how the essential features of Rawls's chosen society of concern shapes the deliberation process behind the veil of ignorance, leading to his principles of justice.

Now we will provide a similar analysis for our society of concern.

OUR SOCIETY OF CONCERN, ITS ESSENTIAL FEATURES, AND OUR FIRST PRINCIPLE OF JUSTICE

Let us begin by describing our society of concern and differentiating it from that of Rawls's. Our society consists of many species of animal used by humans primarily for human benefit. The manner in which humans use these animals causes varying degrees of suffering and most often death.

The ARC, or the individual members thereof, are facilitators of the distribution of the resources meant to help the animals in our society. Members of the ARC are not a part of our society of concern. For, they are not in need of the kind of justice that is sought for the animals within their concern.

Our society is not a cooperative venture for mutual advantage. Indeed the individuals in our society do not stand in any cooperative relationship to one another. Hence, any benefits to be gained by individuals in our society will not be generated by the cooperation of the members of our society. Rather, the animals in our society merely stand to benefit passively through resources possessed by members of the ARC for the purpose of abolition.[9] Moreover, since the purpose is abolition of the use of animals, the animals can expect mutual benefit.

Whereas Rawls's society is a cooperative venture for mutual advantage, then, our society is the target of a charitable venture for the mutual advantage of its members. The following three essential features of our society of concern frame the deliberation process behind the veil of ignorance leading to our two principles of justice:

Feature 1: Society consists of animals who are confined, suffering to varying degrees due to being used by humans, incapable of freeing themselves, and facing an untimely death.

Feature 2: Our society is the target of a charitable venture by the ARC and its resources.

Feature 3: The goal of the charitable venture is mutual advantage among the animals in our society via the abolition of their use.

These features are part of what is known from behind the veil of ignorance. Principles of justice for our society could not be arrived at without knowing these features, just as Rawls's principles of justice could not be arrived at without knowing the essential features that define Rawls's society of concern.

Before employing the veil of ignorance to begin the derivation of our principles of justice, I will provide a brief sketch of what I will know about myself and my role from behind the veil of ignorance. Since I know about the above three features, I know that I will be some kind of sentient animal of concern to the ARC. And I know that I will be living a life that would be better off for me if humans were to refrain from using me. And of course, I also know that it is my duty, from behind the veil of ignorance, to decide on the basic principles for the distribution of ARC resources, and to do so with my own interests taking priority. Knowing these truths behind the veil of ignorance does not give me any type of unfair advantage over others in my society.

I will now adopt the metaphor of being behind a veil of ignorance, and I will reason in accordance with my own self-interest as one who will be a member of our society.[10] From behind the veil of ignorance, although I know that I will be an animal used by humans, I will not know whether I will be a farm animal, a hunted animal, a research animal, a zoo animal, and so on. Knowing any of these particular facts about myself would allow me to choose rules that give more weight to my preferences as a part of this society, which would not be fair. Also, I will follow Rawls and assume that I do not know the probability that I will fall into any category of animal.[11]

Now, since I do not know what kind of animal I will be, I would not want any category of animal to be abandoned by the chosen distribution. It would be rare for it to be rational to will one's own abandonment when faced with the kinds of suffering and death we are talking about, suffering and death I know about from feature 1 of our society. This is how I arrive at what I will call the principle of non-abandonment. The principle of non-abandonment guards against leaving out any category of animal that suffers due to being used by humans, and it is certainly in my self-interest to endorse such a principle. Moreover, I take this principle of non-abandonment to be intuitively strong enough to be considered prima facie nonnegotiable.

I would also understand that feature 3, the requirement of mutual benefit, implies a principle of non-abandonment.

DETERMINING OUR SECOND PRINCIPLE OF JUSTICE

Our principle of non-abandonment stipulates that no category of animal be abandoned. However, it does not provide us with a guide for how to distribute the available resources among the different categories of animal. In considering this issue, I would understand that since I cannot abandon any category of animal, I must choose either equal distribution or an unequal distribution that precludes abandonment.

By "equal distribution" I would mean the same quantity of resources per category of animal. I would not define "equal distribution" as equal resources per animal because I would fear ending up in a category containing a small number of animals, which would lead to fewer resources for my category.

Now, initially, in attempting to decide between equal and unequal distribution I would be worried about my category of animal ending up with fewer resources than animals from other categories. So I would take seriously the option of equal distribution. However, I would be willing to opt for unequal distribution if a case could be made that unequal distribution would be preferable to equal distribution.[12]

Hence, what is left undecided is whether there is a framework for inequalities in resource distribution that would be chosen from behind the veil of ignorance and from which we can extract a principle of distribution. I have identified three frameworks for unequal distribution which are candidates for being competitors with equal distribution. I will call them "the utilitarian framework," "the libertarian framework," and "the severity of suffering framework."

The Utilitarian Framework

By "the utilitarian framework" I mean either the classical variety of utilitarianism that we get from Bentham or Mill[13] or the preference utilitarianism[14] of Peter Singer. The two positions we will consider, respectively, are grounded in these forms of utilitarianism. And since the difference between these two kinds of utilitarianism will not make a difference to the analysis that follows, I will mostly use the classical language of happiness and suffering when referencing what the utilitarian sums over.[15]

Recently, a utilitarian argument of the classical kind has been made that those who care about the plight of animals against human-caused suffering and death ought to concentrate only on abolishing factory farming by convincing people to stop eating animals.[16] Matt Ball (2006), for example, states: "Convincing just one person to change his or her diet can spare more animals than have been saved by most of the high profile campaigns against animal research, fur, and circuses." Ball (2006) concludes: "It is clear that, if we want to maximize the good we accomplish for the animals, expanding the boycott of factory farms through the promotion of vegetarianism is the best use of our limited time and resources."

Erik Marcus also argues for focusing on abolishing factory farming. He states, "There can be no doubt that the effort to eliminate cruelty to animals should focus on agriculture" (Marcus, 2005). He calls for a "dismantlement" movement to stop animal agriculture (Marcus, 2005).

Farm animals are by far the most numerous of the animals used and killed for human purposes. Marcus (2005) calculates that 97 percent of all animals killed in the United States are farm animals. The other 3 percent are killed during animal research, hunting, fur farming, and shelter work. Marcus (2005) concludes: "Farmed animals therefore deserve priority, and arguments made on their behalf should not be weakened by rhetoric pertaining to hunting, medical research, or companion animals."

The argument, then, is that in terms of quantity of suffering, farm animals' cumulative suffering greatly outweighs the cumulative suffering of all other animals that are mistreated and/or killed for use by humans. Concentrating on convincing people to stop eating animals, it is argued, will therefore have the greatest and quickest impact on the total quantity of animal suffering in the world. Human resources are therefore best directed toward this goal.

Philosopher Peter Singer (2002), a preference utilitarian, seems to agree. He argues that the United States "lags . . . far behind Europe in its protection of animals." He attributes this to the U.S. animal activists' focus on animals used in research rather than on animals raised for food, while pointing out that there are at least 100 farm animals to every research animal (Singer, 2002). Singer implies that had the U.S. movement focused on farm animals, more suffering could have been addressed more quickly, as has happened in Europe.

Notice that the position advocated here is not simply that we ought to focus more resources toward abolishing factory farming, but that we ought to focus *all* resources toward abolishing factory farming. According to this view, it would be morally negligent to waste resources on any other cause when abolishing factory farming is the most efficient way to diminish overall suffering.

It is true that the sheer numbers make the utilitarian strategy compelling from a moral perspective generally. However, although the utilitarian strategy may be compelling on this level, it calls for all resources to be designated for only farm animals. It therefore ostensibly violates our non-abandonment principle. For, assuming that 10 billion animals are killed for food in the United States per year and that there is 1 research animal for every 100 farm animals, focusing only on farm animals essentially abandons 100 million research animals, in the United States alone, every year, not to mention the additional 200+ million animals abandoned in the United States from all other categories (if Marcus's percentages are correct). Therefore, it fails as a candidate framework for unequal distribution.

Notice that the utilitarian arguments considered above exemplify a problem common to utilitarian views, namely that, as Rawls (1999) argues, "utilitarianism does not take seriously the distinction between persons." For example, speaking of utilitarian views generally, Rawls (1999) states: "There is no reason in principle . . . why the violation of the liberty of the few might not be made right by the greater good shared by the many." In other words utilitarian views have a problem with protecting the individual from the requirement of maximizing overall happiness or of maximizing the satisfaction of preferences. This is, at bottom, why the utilitarian arguments made by Ball, Marcus, and Singer fail to satisfy our principle 1.

The Libertarian Framework

Perhaps ARC individuals ought to give to the animal causes based on their individual empathies. Some individuals are very devoted to halting animal research, some to spreading veganism, some to rescuing stray and abandoned dogs, and so on. The libertarian idea here would be not to constrain individuals of the ARC in their decisions about which animals to help.

The argument for distribution of resources according to where empathies lie is that what people are passionate about is the best indicator of how hard-working and effective they will be in meeting their goals. It is also an indication of what percentage of their income they will devote toward their cause. Since they will be less effective working on other causes, the best thing to do, it seems, is to match causes up with those who are most passionate about them. There will certainly be inequalities that arise in this framework for distribution, but, the libertarian argument goes, it would be justified by the benefit to all animal causes due to each cause receiving help only from those most devoted to it.

The way that this libertarian framework might be chosen from behind the veil of ignorance is if those receiving fewer resources could always have the highest possible impact on helping their animal cause, merely due to there being the highest possible average passion among their fellow activists. In other words, the libertarian framework might be chosen from behind the veil of ignorance if highest average passion were the only thing that mattered to effectiveness. However, I think that this is clearly not true.

Suppose that the libertarian framework results in group #1 having many more people working on its cause than group #2. Now, adherence to this framework would result in group #1 and group #2 each having the highest possible average passion among their members for their respective causes. Moving any people from group #1 to group #2, for example, would result in a lowering of the average passion in group #2.

Suppose now that group #1 also has much more *money* than group #2. Group #1 then would not only have the highest possible average passion among their fellow activists, as does group #2, but group #1 would also have more people and money. Under these circumstances group #1, in ordinary cases, will have an advantage over group #2, in terms of effectiveness, even though they each have the highest possible average passion in their groups.

Moreover, group #1's and group #2's advantage and disadvantage respectively would be due to circumstances beyond the control of animals in the two groups. More people just happened to be more inclined to help, and fund more heavily, group #1 than group #2. Perhaps group #1's animals are aesthetically more pleasing than group #2's. The disadvantage to animals in group #2, nevertheless, would be due to accidents of their nature and circumstances. And recall that such accidents ought not to place individuals at an automatic disadvantage in their society. This framework would not be chosen from behind the veil of ignorance because one would be worried about ending up in groups like group #2.

Also, the libertarian framework leaves open the possibility that the members of some categories of animal—because they are not the recipients of empathy—be abandoned, which would violate our non-abandonment principle—another reason to reject the libertarian framework.

The Severity-of-Suffering Framework

Behind the veil of ignorance, an individual should be very concerned that one does not know which kind of animal one might turn out to be. The reason for this concern would be due to the differences in the severity of suffering among different categories of animal. On average, animals in zoos probably do not suffer as much as factory-farmed animals. Some research animals probably suffer more than some factory-farmed animals. Animals hunted for their fur probably suffer less over the course of their lives than those raised for their fur.

Due to these differences in severity of suffering, I maintain that we ought to categorize animals in terms of severity of suffering for the purpose of determining how to distribute ARC resources. Consider that one would understand from behind the veil of ignorance that one could end up being an individual whose suffering is among the highest (the least advantaged), a being whose suffering is among the lowest (the most advantaged), or a being whose suffering is somewhere in between. And one would fear ending up in a least advantaged category more than one would fear ending up in a most advantaged category. The worse a category, then, the more one would desire help, and the more help one would desire, if one were to end up in that category. From behind the veil of ignorance, then, one would want the animals whose suffering is among the highest to be provided more resources than those whose suffering is among the lowest.

Given the uncertainty about where one might end up along the continuum of suffering, then, it would be rational for one to accept a distribution of resources which is proportional to the severity of suffering of the animals populating the different categories. On this severity-of-suffering framework all those who are in need are provided for in proportion to that need.

For these reasons, I maintain that the severity-of-suffering framework would be chosen from behind the veil of ignorance. We can now state the two principles of justice for the distribution of the ARC's resources:

1. The Non-Abandonment Principle: No category of animal ought to be abandoned by the chosen resource distribution framework, and
2. The Severity-of-Suffering Principle: Inequalities in resource distribution are to be arranged so that they are distributed in direct proportion to the severity of suffering of the category of animal.[17]

Now I want to briefly summarize how the essential features of our society led us to these principles. Knowledge of feature 1 or 3 of our society is enough to convince those behind the veil of ignorance that we require a non-abandonment principle. Feature 1 tells me that I will be confined and suffering and unable to escape on my own. This gives me the self-interested motivation to choose a principle that precludes my own abandonment. Feature 3 tells me that mutual benefit is the goal. This directly implies that no category of animal be abandoned. Hence features 1 and 3 are each sufficient to conclude that a non-abandonment principle is a rational choice for our society of concern.

As for the severity-of-suffering principle, this principle can be seen to arise primarily out of features 1 and 2 of our society of concern. Knowing that the confinement and suffering comes in degrees and knowing that resources come from a charitable source, one would base distribution on need, where the greater the severity of suffering, the greater the need, the greater the resource allocation.

EXPLANATION OF THE DIFFERENCES BETWEEN RAWLS'S PRINCIPLE AND OUR PRINCIPLES

We have seen that the distinct essential features of Rawls's and our societies lead to different principles of justice. Here I want to explain why it is that they come out different. I do this by comparing the two sets of societal features corresponding to the two sets of principles.

The principle of non-abandonment is analogous to Rawls's (1999) liberty principle only in the sense that they both acknowledge nonnegotiable individual rights with respect to sharing in society's resources.[18] However, whereas Rawls's liberty principle grants individuals liberty rights to pursue society's resources relatively unencumbered, our principle of non-abandonment grants positive rights to individuals to receive some of the ARC's resources. Our first principles differ here due to the differences in the corresponding features 1 that define our societies of concern. Rawls's society consists of agents who act to further their interests, which implies the need for liberty rights. Our society is constituted by animals who are in need of assistance because they cannot act to further their interests, which implies the need for positive rights.

Rawls's principle 1 and our principle 1 differ also because our principle 1 plays the role of securing the goal of feature 3, mutual benefit, whereas Rawls's principle 2a is what achieves mutual benefit by constraining his principle 2b. This is because our distribution is based on need which means we require a principle that achieves mutual advantage by stipulating that no one in need be abandoned. In contrast Rawls's distribution is based on what one can acquire employing one's liberty. Hence, to achieve mutual advantage in this case, one needs to reign in what can be achieved with one's liberty by mitigating the effect of natural and social contingencies on what can be achieved.

There are two differences between our principle 2 and Rawls's principle 2. They can both be explained by a difference in the features 2 of Rawls's and our societies of concern.

The first difference is that whereas Rawls's principle 2a guarantees a minimum of benefits for the least advantaged, our principle 2 guarantees a maximum of benefits for the least advantaged. This is because in Rawls's society, resources accrue to individuals according to what they can acquire employing their liberty in cooperation with others, whereas our society's benefits accrue to individuals according to need.

To see this, first consider that in Rawls's society, some people are the least advantaged due to circumstances beyond their control. These are disadvantages due to shortcomings in their nature or social circumstances. And these disadvantages inhibit their ability to acquire benefits in cooperation with others. Now recall that principle 2a requires that economic and social inequalities be to the greatest benefit of the least advantaged. The greatest benefit is the point where any increase in these inequalities would decrease, rather than increase, benefits to the least advantaged (Rawls, 1999). This point, then, marks a minimum requirement for justice. Principle 2a therefore acts as a safety net (a minimum) to mitigate fortuitous circumstances responsible for the above kinds of disadvantages.

Now, our society seeks to help individuals in need due also to circumstances beyond their control, circumstances largely resulting from the biases of individuals in the ARC in favor of helping some kinds of animal over others. According to our principle 2, where need is the requirement for assistance, those who have a greater need will receive more help. In contrast to Rawls's society, the least advantaged in our society receive a maximum of benefits.

It is also worth noting that in Rawls's society, the most advantaged receive a maximum limit to their benefits. For, they cannot acquire more benefits if that acquisition would bring down the advantages accruing to the least advantaged. In contrast, the most advantaged in our society receive some minimum limit to their benefits due to the non-abandonment principle.

The second difference between our principle 2 and Rawls's also reflects a difference in our societies of concern, again a difference between Rawls's and our feature 2. Our principle 2 has no part "b" similar to Rawls's because in our society of concern resources are neither generated by nor distributed according to cooperation among employed individuals. Our society is the target of a charitable venture. Hence, the notion of fair equality of opportunity has no role to play in the production or distribution of resources in our society.[19]

CONCLUDING REMARKS

One implication of distributing resources according to severity of suffering is that the differences in suffering will not only cut across our typical animal activist categories but also exist within typical categories. For example, consider the category of hunted animals. Animals killed quickly by experienced bow-hunters certainly do not suffer as much as those who are caught in hunters' leg traps and lie dying for several days. Consider the category of research animals. Pigs or primates who are used for psychological experiments almost certainly do not suffer as much as those used in head trauma experiments. According to Erik Marcus (2005), chickens have the worst overall life compared to any other kind of animal in the farm animal category.

What this means is that if we categorize animals in terms of severity of suffering, we are going to be concerned about nontypical categories of animals. We might find that homeless cats and zoo animals are at the same severity of suffering and belong in the same category. We might find that animals subject to psychological experiments are in the same category as circus animals in terms of severity of suffering. We might find that battery-cage layer hens are in the same category as animals used in certain neurophysiological medical experiments.

Due to this need to determine different levels of suffering within and across typical categories of animal, I do think that the severity-of-suffering framework is in need of more development. Primarily, we would need a rough ontology of severity of suffering. And although it may be difficult to determine a precise just distribution based on severity of suffering, this does not mean that activists cannot work toward approximating such a distribution. And I think that our two principles will yield a result that is superior to any framework that abandons even one category of animal or distributes resources according to the biases of individuals in the ARC.

This chapter was previously published: Alan C. Clune, "Rawls and the Distribution of Human Resources By Those in the Animal Rights Community," International Journal of Applied Philosophy, Volume 28, Issue 2, Fall 2014, pp. 251–266. DOI: 10.5840/ijap2014112431

NOTES

1. By "animals" I will mean sentient nonhuman animals.
2. On p. 4 and p. 109, we see that Rawls (1999) holds that cooperative societies yield greater benefits.
3. See Rawls (1999), pp. 109–112. See also p. 442 where Rawls acknowledges behind the veil of ignorance one knows that one will be a moral person.
4. This is the point at which the increasing advantage for those better situated starts to bring down the admittedly lesser advantage that had been accruing for the least advantaged. See Rawls (1999), pp. 65–70.
5. The just savings principle accounts for what we owe future generations. See Rawls (1999), pp. 251–258.
6. See Rawls (1999), pp. 86–93. Here he explains why principle 2a precludes principle 2b implying a meritocracy in the distribution of benefits from cooperation.
7. Rawls calls the difference principle "a principle of mutual benefit." See Rawls (1999), p. 88 and p. 89.

8. Note that principle 1 precludes the form of government from being a meritocracy, but principle 2a precludes a primarily meritocratic distribution of men's shares (their wages, for example) in the benefits derived from social cooperation. One's system of government could be non-meritocratic, but without principle 2a, principle 2b could still lead to a meritocratic distribution of shares in wages by default.

9. I intend for any conclusion I draw to be applicable to individual ARC members or groups of ARC members.

10. Employing the veil of ignorance requires only one individual. See Rawls's (1999) discussion on pp. 119–120.

11. See Rawls's (1999) discussion on pp. 133–134. Rawls argues that it is rational to avoid worst case scenarios for oneself, independently of how improbable they may be.

12. Rawls (1999) makes this point when arguing for his difference principle on pp. 130–131. See also p. 54.

13. This is the view that one ought to act in such a way as to maximize overall happiness or minimize overall suffering.

14. Preference utilitarianism is the view that one ought to act such as to maximize the number of satisfied preferences.

15. One reason Singer holds preference utilitarianism is because he wants to take account of a difference between humans and animals when it comes to maximizing happiness, namely, that humans can have preferences that reach into the future whereas animals presumably cannot formulate such preferences. Since our community includes only animals, the language of classical utilitarianism is sufficient for our purposes.

16. Although not explicitly stated, from the language that is used, this appears to be a classical utilitarian argument. However, preference utilitarianism would arguably lead to the same conclusions—at least when applied only to animals.

17. Principles along these lines, I believe, would be applicable to human charitable ventures as well.

18. Excepting extraordinary circumstances, Rawls (1999), p. 132.

19. There is no need to take account of future generations of our society because the ARC's resources will continue as long as there are animal injustices.

REFERENCES

Ball, M. (2006). Living and working in defense of animals. In P. Singer (Ed.) *In defense of animals: The second wave* (pp. 181–186). Blackwell Publishing.

Francione, G. (2008). *Animals as persons: Essays on the abolition of animal exploitation*. Columbia University Press.

Marcus, E. (2005). *Meat market*. Brio Press.

Rawls, J. (1999). *A theory of justice*, Revised Edition. Belknap Press of Harvard University Press.

Regan, T. (1983). *The case for animal rights*. University of California Press.

Rowlands, M. (2009). *Animal rights: Moral theory and practice*, second edition. Palgrave, Macmillan.

Singer, P. (2002). *Animal liberation*. Harper Collins.

Chapter 3

The Fabric of Life

Technology, Ideology, and the Environmental Impact of Clothing

Juneko J. Robinson

As a graduate of the now-defunct Yodogawa School of Western Style Dress in Osaka, Japan, my mother's teenaged ambition was to open a dress shop featuring her own designs. That dream never came to pass, but my mother never relinquished her love of clothing. When she died nine years ago, she left behind a substantial amount of clothing and accessories all in near-mint condition thanks to a combination of her fashion-school training, naturally occurring fastidiousness, and the fact that she once worked at Frankie Yee's, a long-gone Chinese laundry in Novato, California, the town of my birth. To comb through the contents of her closets and dresser drawers was to be confronted not only with a cavalcade of fashion but with every clothing-specific memory I'd ever had and, as a former aspiring fashion designer, I had a lot of them. Here was the tasteful black dress with matching belt worn only once, to my knowledge, at my brother's funeral back in 1993, the 1980s-era bright fuchsia coat that she rejected in favor of a stolid gray one for her first trip back to Japan in thirty years, because it was "improper" for a woman of her age to dress in such youthful colors, and the groovy black halter dress with neon-lights-at-night white, magenta, and chartreuse geometric patterns that she sewed for my father's fortieth birthday party back in 1971, and in which I thought she was particularly beautiful. This last, I kept.

But one of the oldest pieces of clothing that my sister and I came across as we began the painstaking inventory of our late mother's estate were a pair of skin-tight, just-above-the-knee shorts once improbably known as "Jamaica shorts." A forerunner of 1990s-era "bike shorts," Jamaica shorts were ubiquitous in the 1960s and consisted of vertical stitched front seams running down the thighs and, in this particular case, made of a heavy navy blue polyester double knit of a particularly dense and durable weave that hasn't been manufactured for decades. What was remarkable was the fact that they appeared as if they were still brand new—no fading, no pilling, and still shapely and stretchy—despite the fact that my mother wore these constantly during the warm months of my early childhood. Indeed, I have a very specific memory of her wearing just those shorts while seated at our backyard picnic table during a family barbecue sometime between 1969 and 1973, the years that we lived in that particular house, but they may very well be the same pair that she is wearing in the very first photo taken of me when I was approximately two months old, back in 1966. I took a photo for posterity's sake and then promptly bagged them for donation. The forty-plus year lifespan of my mom's Jamaica shorts speaks volumes about the remarkable quality of many goods that used to be manufactured for people of modest means. But it also represents a troubling legacy bequeathed upon us by way of the "better living through chemistry" model of manufacturing, one whose impact we are only just now beginning to appreciate.

Contrary to current understandings, my mother was not a hoarder. Indeed, any such comparison would be ironic given that the sheer quantity of stuff that the average American owns is more than quadruple that of most people in the mid-twentieth century, thus qualifying most of us as "hoarders" by the standards of the 1950s and 1960s, especially given that, in 1950, the average newly built house was just 983 square feet (compared to 2017's figure of 2,631) and thus had inadequate space for storing the sheer quantity of goods that permeate today's homes (Comen and Sauter, 2019). To the contrary, my mom was a woman who was born in the midst of a global depression, and a survivor of both World War II and a financially challenged marriage, which meant periodic stints on food stamps, and, later, a single, working-turned-middle-class mom. Like most of her generation, she was frugal both out of virtue and necessity. She also had tremendous skill. Having grown up during a time of wartime austerity in a culture that valued traditional "feminine" arts, as well as having trained as a designer, my mother learned how to sew, make dress patterns, knit, embroider, needlepoint, and crochet and used these skills to upholster our home's furniture and to stylishly dress herself and her family, despite our limited income. I remember accompanying my mom to a local dime store, the TG & Y, to pick out notions and fabric, as well as watching her work at the sewing machine. Occasionally, I would be allowed to help. No strangers to the amount of work that went into garment construction, we, as children, learned to value our material possessions and the work that went into making them. Clothing was scrupulously cared for, mended, and altered to last as long as was feasible, which was possible in part because even modestly priced clothing was comparatively well-made. And, when our clothes were beyond mending, fabric, buttons, and zippers were conscientiously culled and re-used for new fashions, curtains, and quilts. The contents of my mother's closet hung as mute witness to a way of life that had, unbeknownst to me, quietly passed.

FAST FASHION

Like most television viewers, I was aware of the emergence in the 2000s of shopping as a publicly avowed hobby. Although there is disagreement over when the term "shopaholic" first entered the lexicon (the *Oxford English Dictionary* states that it first appeared in print in 1984, while *Merriam-Webster* puts it at 1977), it wasn't until the 2000s that I began to routinely see shopping touted in the media as some sort of sociological master status. Indeed, the opening credits of one season of the globetrotting reality competition *The Amazing Race*, introduced one competing female duo walking down the street, smartly dressed, each carrying a large number of designer shopping bags, as if shopping was the most salient characteristic they wanted the world to know about them. However, the extent to which the comparative thriftiness and resourcefulness of earlier generations had been displaced had largely escaped me until I was confronted with the following statistic: according to author Elizabeth Cline (2012), in 2008, Americans bought an average of sixty-four items of clothing per year, a little more than one piece of clothing per week. Apparently, shopping was indeed the new pastime, not only for Americans, and other affluent "First Worlders," but also for a new burgeoning middle class in the industrialized developing world.

Indeed, global consulting firm McKinsey & Company found that worldwide clothing production doubled from 2000 to 2014 (Remy et al., 2016). Leading the way were the new generation of global discount fashion retailers such as fast-fashion pioneer Zara (Spain), H&M (Sweden), Uniqlo (Japan), Gap/Old Navy (U.S.), Topshop (U.K.), and H & A (Belgium-Germany-Netherlands), which spread throughout the developed world as well as in the emerging markets of Brazil, Russia, India, China, and South Africa, that is, BRICS (Remy et al., 2016), and other newly industrialized countries beginning in the early 2000s. With an entire business model premised on cheap clothing, such retailers soon realized that they could stoke insatiable demand by constantly introducing new designs. The old way of doing things collapsed almost overnight. Where there were once two to four seasons per year within the fashion industry, multinational retailers like Zara began offering more than twenty new collections per year (Jenna, 2017), while Topshop introduces an astounding 400 new styles per week on its website (Cline, 2012).

This was due to a confluence of a number of factors: deindustrialization in the First World, rising incomes and industrialization in parts of the developing world, and the concomitant outsourcing of clothing and textile manufacturing from the United States and other nations comprising the "Global North," thus ushering in the twenty-first century's newest version of colonialism. For big, multinational corporations and their investors, business, needless to say, has been very good. However, the corrosive effect of fast fashion has been devastating both culturally and environmentally.

DIRTY LAUNDRY: THE ENVIRONMENTAL IMPACT OF FASHION

Despite our long-standing concern with the environment, the devastation wrought by the fashion industry has, until recently, largely escaped public scrutiny even though textile-making is one of the oldest forms of human production dating back to approximately the sixth millennium (Moulherat et al., 2002) and was one of the driving forces behind the Industrial Revolution. And that environmental impact occurs at every stage throughout the lifecycle of a garment. Although the true environmental impact of the entire supply chain of clothing is too complicated to discuss at length here, the following should be illustrative.

Fashion production adversely impacts water. Although over 70 percent of the Earth is covered in water, only some 2.5 percent of it is fresh and, of that, only about one percent of it is accessible with the rest stored in the form of glaciers and snowfields ("Freshwater Crisis," n.d.). Given the scarcity of fresh water, the current state of clothing production—both in terms of usage and pollutants—is problematic. The most obvious culprits are synthetic fabrics, which are made from nonrenewable and heavily polluting petroleum and coal products. Such fabrics are also nonbiodegradable and release harmful substances into the environment during chemical processing such as dyeing, waterproofing, and flame proofing. Moreover, synthetic fabrics are also shed and dispersed into our waters with laundering and are thus a major cause of microplastics in the environment, particularly in urban areas (Hernandez et al., 2017). Such microplastics are ingested by and kill sea life and likely have adverse health effects on humans. Disturbingly, the ever-popular fleece fabric, initially championed because it can be made from recycled plastic PET bottles, is a major culprit.

Less intuitive, however, is the fact that, although natural fibers such as those derived from plants or animals are biodegradable and renewable, they are not necessarily sustainable. Most of the world's garments are made from cotton and it is the single most important nonfood crop in the world (Interchurch Organisation for Development Cooperation et al., n.d.). A thirsty crop that tends to be grown in arid regions, conventional cotton cultivation, combined with our exploding population growth and increased pressure from agriculture and industry, adds to freshwater scarcity ("Freshwater Crisis," n.d.). In an age of diminishing aquifers, agricultural irrigation is depleting freshwater resources in many areas faster than they can be replenished, which calls into question whether such a precious commodity should be used for textile production at the level currently in demand. For example, the Ogallala Aquifer on the High Plains of the United States is responsible for some 30 percent of all irrigation in the United States and supports one-sixth of the world's grain produce; however, it has long been unable to meet agricultural demands (Frankel, 2018). A portion of that demand stems from the United State's largest cotton producer Texas, as well as that of New Mexico and Kansas (National Cotton Council, 2013). In addition to jeopardizing global food security, aquifer depletion also leads to fish extinction, a situation that serves as a catalyst for biotic homogenization, which in turn renders aquatic communities less resilient (Frankel, 2018). Moreover, because the world's water circulates as a closed system, the fact that cotton production accounts for some 50 percent of insecticide use across the developing world and some twenty-five percent of insecticide use globally (Food and Agriculture Organization of the United Nations and International Cotton Advisory Committee, 2015), contamination from cotton production often renders water unfit for other purposes such as drinking and food cultivation (Fletcher & Grose, 2012).

There are also environmental impacts associated with monoculture agriculture, that is, the repeated planting of the same plant species in the same soil. Although there are some fifty-three species of cotton, only four are widely cultivated (Gotmar et al., n.d.) and over 90 percent of the world's cotton stem from one species (Shim et al., 2018), which decreases biodiversity. Monocultural production also adversely impacts soil through salinization and prevents the regeneration of soil nutrients that would come with rotating different varieties of crops, which, in turn, leads to soil depletion. Finally, reduced genetic variety also leaves crops vulnerable to disease and climate stressors, such as drought, which can result in crop failure.

Recent meta-studies have found that some 97.2 percent of scholarly papers published across an array of scientific disciplines find that climate change is caused by humans (Cook et al., 2016). Although automobiles are usually the first culprits we tend to think of, according to the United Nations Environment Programme (2018), the fashion industry is responsible for ten percent of the world's carbon emissions—more than all international flights and maritime shipping combined. Even natural fiber production and consumption emit carbon: some sixty percent of the energy used in the life cycle of a single cotton t-shirt comes from post-purchase laundering—far greater than that expended in its transportation from producer to consumer

(Claudio, 2007). However, according to some estimates, the movement from conventional to organic cotton alone could reduce the industry's contribution to global warming by some forty percent ("Pulse of the Fashion Industry, 2018," n.d.).

Finally, there is the staggering number of clothing items that are discarded annually. The rock bottom pricing of fast fashion means that shoppers no longer view clothing as an investment, even as they're buying 60 percent more clothing in 2014 than they were in 2000 (Remy et al., 2016). Moreover, poorer quality construction means that apparel simply doesn't hold up to repeated wear and washings like it used to. As a result, the amount of clothing articles discarded annually doubled from some seven million to fourteen million in less than twenty years to an average of eighty pounds per person, per year (Wicker, 2016). From 1960 to 1970, the likely birthdate of my mother's Jamaica shorts, the U.S. Environmental Protection Agency (EPA) estimates that between 1,360 and 1,620 tons of clothing were generated in the United States with the vast majority being sent to municipal landfills at the end of their lifespans (Environmental Protection Agency, n.d.). During that period, only some fifty tons were recycled and none were incinerated (Environmental Protection Agency, n.d.). By 2017, those numbers had ballooned to some 12,800 tons of clothing generated, resulting in 8,900 being sent to landfill, 2,160 tons being incinerated (with energy recovery through landfill gas capture), and a mere 1,740 tons being recycled (Environmental Protection Agency, n.d.). How on earth did we get here?

TECHNOLOGICAL THINKING

Underlying such destructiveness is a way of thinking and relating to the world that has led to the very situation in which we currently find ourselves, and which colors any likely political responses to this environmental crisis. Proponents of technology often encourage the belief that technology is value-neutral; that we have only good things to gain and that whatever is lost was probably not worthwhile keeping. Insofar as technology allows us to develop and use tools to engage in activities quickly, efficiently, and successfully, it is good, but otherwise it is morally neutral, containing no inherent values in and of itself. The philosopher Martin Heidegger thought otherwise.

For Heidegger there are three aspects to modern technology. First, there are the techniques, devices, systems, and production processes associated with industrialism; second, there is the rationalist, scientific, commercialist, utilitarian, anthropocentric, secular worldview usually associated with modernity, and third there is the "contemporary *mode of understanding or disclosing things* which makes possible both industrial production processes and the modernist worldview" (Zimmerman, 1990). It is these last two that are most crucial.

According to Heidegger, ours is the technological age, not because of any particular technological development, but because technology demands a one-dimensional way of disclosing the world and everything in it as raw material to be used as fodder for the acquisition of power. This ability of technological thinking to reveal everything in nature as stored energy and subsume everything unto itself, independent of our intentions, is what Heidegger referred to as "enframing" (Heidegger, 1993). This technological enframing predisposes us to view everything, including human beings and other living entities, in terms of their use or function. Echoing this, cultural critic Neil Postman has similarly noted that "technopoly"—that is, the culture of technology—creates its own imperatives by

> eliminat[ing] alternatives to itself precisely the way Aldous Huxley outlined in *Brave New World*. It does not make them illegal. It does not make them immoral. It does not even make them unpopular. It makes them invisible and therefore irrelevant. And it does so by redefining what we mean by religion, by art, by family, by politics, by history, by truth, by privacy, by intelligence, so that our definitions fit its new requirements. Technopoly, in other words, is totalitarian technology. (Postman, 1992)

Technological values such as instrumentalism, speed, quantity, efficiency, calculative thinking, the stockpiling of surplus, and the creation of waste have allowed us to further develop specific techniques and technologies that leverage such values. Indeed, technological thinking encourages instrumental reasoning that is, a kind of rationality whereby the most economical means to an end are calculated (Taylor, 1991) so as to effect gains in speed, quantity, efficiency, and so on. Hence, technology constitutes the foundational nature of our age, providing the lens through which all of the world's entities are viewed, pervading every aspect of our world, and revealing it as a totality of beings that can be explained, controlled, used, consumed, stockpiled, and leveraged as a means of achieving, growing, and maximizing power, which is why Heidegger concluded that the essence of technology is not anything technological (Heidegger, 1993).

The problem with technological enframing is that it encourages shallow, superficial ways of living. It narrows our attention to things, rather than to other beings, and relegates us to questions concerning usage and stockpiling, rather than to those of existence as such, and it does so by encouraging behavior such as the relentless seeking of novelty, the amassing of things and facts for their own sake in order to briefly "see" them and then quickly move on (Fandozzi, 1982), and engage in what Heidegger referred to as "idle chatter," which never dwells on any topic at length, but touches on as many things as possible. Such ways of being encourage us to avoid probing too deeply or inquiring too profoundly into the world around us and are diametrically opposed to living deeply in recognition of the world as a single unity.

These intuitions about the nature of our current age are shared by some who are proponents of deep ecology, an environmental philosophy that holds that living beings have their own inherent worth above and beyond any utility to human beings. Many agree with Heidegger that any meaningful solution to our environmental crisis necessarily involves an ontological shift away from our current anthropocentric, dualistic, and utilitarian understandings of nature, which only disclose the world as raw material for human ends (Zimmerman, 1993). At the same time, however, other deep ecologists reject Heidegger's philosophy, partly due to his anti-naturalism stemming from his denial that humans are animals and thus a part of the natural world, and partly because they suspect him of anthropocentrism, despite his claims to the contrary (Zimmerman, 1993). Ironically, this tension between humanists (although Heidegger would resent his classification as such) and environmentalism itself stems from a long history of dualism in Western thought. As Matthew Ally has observed, "Whether it is taken as a passive and mute background to higher-order human concerns, or some wild otherness to be tamed, or as repository and toolkit for the satisfaction of human needs, a strong thread of anti-environmentalism sentiment runs through much of the humanist tradition" (Ally, 2017). At the same time, "A strong thread of anti-human sentiment runs through much of the environmentalist tradition, sometimes even despite [their] best efforts to resist or reject it" (Ally, 2017), a subject to which we'll briefly touch on below.

MACHINE AGE IDEOLOGIES

Such philosophical musings about technology and the relationship of humans to nature may seem far and above the nitty gritty of politics and yet modern political theories incorporate much of the same productionist metaphysics first articulated by Heidegger. For example, seventeenth-century philosopher Francis Bacon, an early proponent of what would come to be recognized as a nascent form of the scientific method, encouraged us to think in terms of "operations," rather than abstract truths. Only through a concern for utility could one apprehend the reality of existence and determine the proper course of action and, in doing so, enable science and its technological offspring to improve the human condition by allowing us to control nature to our benefit. With his new methodology, Bacon declared that the end of all knowledge should be a "discovery of all operations and possibilities of operations from immortality to the meanest mechanical practice" (Klein, 2016). Later, the philosopher Thomas Hobbes developed an inductive "science of politics" whereby human nature consisted of atom-like individuals whose natural tendencies toward collision (i.e., "the state of nature" in which every man warred against every other man) could be legitimately controlled by government once they surmised that being subject to the Leviathan's monopolization of power was far preferable to the and nasty, poor, brutish, and short-lived chaos of freedom. Thus, instrumental reasoning, such as cost-benefit analyses, came be seen as applicable to human fulfillment, politics and social engineering as it was to science, technology, business, and economics.

Of course, political theories are only one aspect of political culture. Political culture consists of the entire gamut of "attitudes, beliefs and sentiments that give order and meaning to a political process and which provide the underlying assumptions and rules that govern behavior in the political system" (Aronoff, 2001). To that, we must add ideologies. Like many other aspects of Heidegger's "productionist metaphysics," political ideologies are also distinctively modern because they themselves represent technological thinking as applied to politics. Regardless of where on the political spectrum a particular ideology lays, each expresses an optimism about the human ability to engineer desired social outcomes. As with other forms of technological thinking, ideologies determine which social facts are revealed as relevant and evaluated as social goods (legitimation), subsume all or most human activities under the rubric of politics (integration), provide a master status (socialization), determine policy goals and identify who or what is helpful or detrimental to them (classifying), provide a general map for interpreting the world (simplification), and develop social policy, and the appropriate courses of action for attaining those policy goals (action-orientation) (Freeden, 1996). Finally, insofar as ideologies engage in calculative social cost-benefit analysis and employ instrumental means for attaining, stockpiling, and increasing power, they evince a technological view about how to leverage human collective behavior within a political economy. The contingent nature of the ideologically driven modern world stands in marked contrast to the practices and beliefs of the ancient or medieval world in which people largely accepted their place within a seemingly immutable and natural hierarchy, which was part of a Great Chain of Being. That modernist optimism about the power of human capability exists regardless of whether one adopts hopeful liberalism, cautious conservatism, the communal sentiments of socialism or one of the many new fragmented and "hybridized" "-isms."

Part and parcel of technological enframing is the economic system of capitalism. Over time, capitalist practices have been reinforced by theories about the inherent nature of human beings and *Homo faber* (the creative tool-making man) was transformed into *Homo economicus*. However, once it became a set of normative practices that not only influenced social policy but extended itself to areas of life that had previously been beyond its reach, capitalism arguably became a political ideology. Hence, although it is not usually discussed explicitly as such, capitalism is as much an ideology in the Gramscian sense of "an intellectual, moral, economic, and political unity of aims with the semblance of universality," as it is an economic system (Freeden, 1996). Rather than the Great Chain of Being, we are now inextricably linked to the Great Global Supply Chain. Indeed, because capitalism necessarily has a political dimension to it, it also performs the same functions as ideology by legitimizing, integrating, socializing, ordering, simplifying, and inducing particular forms of action. Hence, it provides us with social meaning and values, directs forms of governance, and develops social policy at every level of government from the local to the international.

As with technological thinking in general, capitalism's true power lays with its ability to absorb everything within its ambit and force the world and its entities to appear in a rather one-dimensional way, namely, as commodities to be bought and sold, all while making it appear as if this arrangement is both natural and inevitable (Appleby, 2010). "Societies that are resistant to capitalist ways today appear unnatural," historian Joyce Appleby writes, despite the fact that it was its western European creators who initially deviated from the global norm (Appleby, 2010). Thus, it is with capitalism that a "new way of establishing political order emerged" (Appleby, 2010). In turn, "People reversed how they looked at the past and the future," according to Appleby. "They reconceived human nature" (Appleby, 2010).

THE CONSUMPTION OF CLOTHING

Beads and colored ochre for self-adornment are among the earliest known human artifacts. These so-called "behavioral Bs," that is, "blades, beads, burials, bone-making, and beauty" (Calvin, 2004) represent some of the earliest examples of what paleontologists refer to as "behavioral modernity," that period in which early humans around the globe began to systematically engage in behavior that would mark them as distinct from their predecessors and lesser-developed contemporaries.[1] Echoing this, *The Oxford English Dictionary* states that the earliest uses of the English verb to "fashion" meaning "to give shape to material or immaterial objects," dates back to the early 1400s, and not until the early 1600s does the term "fashion" come to be identified with one's mode of dress.[2] Hence, fashion, that is, the material, behavioral manifestations of personal appearance is, itself, a form of fashioning our very beings as culturally embedded human beings and indicates the kind of deep existential relationship that we have toward the world and the objects we adorn ourselves with.

Apparel is one of the primary vehicles through which we express our fundamental comportment toward the world, including our lifestyles, values, roles, status, character, and social affiliation and represents an important aspect of tradition and ritual. However, the fact that clothing and/or self-adornment are universal and so deeply imbedded in virtually every aspect of our lives is precisely what rendered them so vulnerable to appropriation and exploitation in the first place. Indeed, agricultural goods related to textile production were among the first to be commodified and were responsible for many of the key technological innovations, such as the loom, the cotton gin, and the sewing machine that led to the Industrial Revolution.

Unfortunately, with technological/capitalist enframing, it is we who are now being fashioned to fit into the global fashion trade, either as disposable workers or consumers of an industry increasingly known for its despoiling of the environment, human rights violations, shoddy craftsmanship, and the promulgation of a global uniform material culture that privileges the aesthetics of the global north and destroys indigenous crafts and local manufacturing.

Like any ideology, the logic of fast fashion covers over its own internal contradictions: as Elizabeth Cline notes, we buy dozens of the same shoes or multiple colors of the same top at bargain basement prices and pride ourselves on our thriftiness, but then wear them once or twice (if at all) and throw them in the garbage, donate them or banish them to the back of our overstuffed closets when we tire of them just weeks later, which hardly seems like a rational use of our hard-earned money even within the context of capitalism, especially considering that real wages in the United States have remained stagnant for decades (Desilver, 2018). Indeed, here in the United States alone, we are now buying and hoarding roughly 20 billion garments per year according to ten-year-old statistics (Cline, 2012). This is echoed in the fact that the average sized master bedroom closet in newly built homes have increased to about six feet by eight feet, a size, more typical of a child's bedroom forty years ago (Cline, 2012), and the fastest growing real estate sector in the United States is self-storage (Fletcher & Grose, 2012). "My sister will pay $400 a month to drive a nice car," Cline writes, "but don't try to charge her . . . more than $40 for a dress. I've seen guys in local coffee shop working on $1,800 Apple laptops and wearing $10 Walmart shoes" (Cline, 2012). It's not that most of us can't afford to pay more for fashion. According to Cline, we just don't see any reason to.

The producers of fast fashion see us, not as symbolic, meaning-conferring individuals who select our dress out of a deeper sense of connection to rituals and other cultural practices that are part and parcel of deeply held values, but rather as mere consumers. Within technological enframing, clothing is simply a cheap commodity with little perceived value, bought in bulk, thereby generating profits for its investors, and then rapidly discarded (or deliberately destroyed by manufacturers, so as to protect intellectual property rights and brand value) (Ferrier, 2018). Indeed, with fast fashion, the particular garment consumed is immaterial. Rather, it is the very act of purchasing alone that confers value, not the object obtained.

THE WORLDING FUNCTION OF CLOTHING

It wasn't always this way. In my mother's generation, clothing was well-made and occupied a greater portion of the household budget. Attaining and maintaining them required more conscious effort. People saved their money for special occasion dresses. Craftsmanship was both expected and widely appreciated. Differences in regional, national, and ethnic dress existed to a far greater degree in a past not known for its celebration of diversity. We were attuned to the changing of the seasons and rituals of varying formality and dressed accordingly. We lovingly cared for our clothing and either passed them down, transformed them, or converted them to other uses. And, when our desires outstripped our budgets, we either made do or made our own. Our old ways of interacting with clothing were arguably more organic, less frenetic. However, many of these practices have since been eclipsed by the imperatives of fast fashion, which have reduced the manifold of our unique and complex clothing cultures to throw away products of consumption in which the ritualized acts of buying and stockpiling are more important than the actual wearing of clothing. It is killing us.

However, there is an alternative way of viewing our relationship to clothing. Heidegger draws a distinction between objects that are present-to-hand (*vorhanden*), which are viewed instrumentally and in a disinterested fashion (Heidegger, 1967), and those that are ready-to-hand (*zuhanden*), that is, objects that we use as if they were organic extensions of ourselves (Heidegger, 1967). In addition, apparel such as shoes are "equipment" (Heidegger, 1993). Equipment is intermediate between mere things and works of art because they are produced by the human hand and can be interpreted through what Heidegger termed "worlding," that is, their ability to evoke the worlds, rituals, and systems of belief that produced and gave meaning to those artifacts. To the extent that we reflect upon equipment, as we do when regarding Vincent Van Gogh's famous painting of peasant shoes, we transform it into art. However, Van Gogh's painting isn't art because it's an object. It is art because it encourages us to meditate on the world of the undepicted peasant who wore them, worked in them, and lived in them; however, unreflectively her own wearing of the shoes was at the time. Hence, for Heidegger, art's specialness stems from its ability to "say something other than what the mere thing itself is" (Heidegger, 1993).

Such evocation can also occur with clothing, particularly when it is animated through wearing. Philosopher Hélène Cixous (1994) has written of the sensuous beauty of wearing a garment that exhibits the "continuity between world, body, hand, garment"; however, such reactions are not limited to philosophers. When Cline tried on a friend's exquisitely designed blazer, she described the difference in quality between it and anything she had ever worn as immediate, "visceral and obvious" (Cline, 2012). Well-crafted clothing is designed and constructed with the total person in mind: how it looks, how it feels, how it moves with the body, where it will be worn, and how it enhances and expresses the values (aesthetic and otherwise) of its wearer. From its conjurings of the farmer who coaxed its nascent plantings from a particular place on earth, the spinner who transformed plant into fiber, the weaver who interlaced the threads into fabric, the artist who imagined its form, the skilled craftsperson who constructed it, to the people who populated the rituals in which our clothed bodies would act, ready-to-hand fashion, as Cixous has noted, connects us to the world around us. It is precisely this type of re-engagement that is needed in order to stem the spiritual and environmental destruction of fast fashion.

Granted, for some, an ideal situation might be one in which consumers adopt a purely "utilitarian" relationship to clothing, assuming that it is even possible. However, green policies that ignore the social and symbolic aspects of consumption are unlikely to alter consumer behavior because they fail to take into account how it facilitates interpersonal interactions and provides a sense of personal identity and worth, as well as a means of creativity (Barnett et al., 2005). Thus, any successful program of reform must take into account the deeper reasons for acquisitiveness with an eye toward channeling them to re-imagine a participatory dress culture that does not depend upon continually buying, stockpiling, and discarding.

ENVIRONMENTAL FASHION

Taking inspiration from the slow food movement, an emerging small-scale fashion industry has sought to promote a different view of fashion; one that questions fashion's emphasis on image and "the new" over making and maintaining actual garments, while seeking to preserve local aesthetics, crafting traditions, and agricultural diversity (Fletcher & Grose, 2012). However, sustainable fashion scholars Kate Fletcher and Lynda Grose caution that, although they have identified a number of ways that the fashion industry can reduce its environmental impact, such innovations are not a substitute for drastically downsizing the output of an industry that typically overproduces by 30–40 percent per season (Magnusdottir, 2020). Rather, any changes should take their cue from nature's strategies, which optimize symbiotic relationships that operate in a dynamic equilibrium (Fletcher & Grose, 2012). Thus, while the development of renewable fibers should certainly be encouraged over nonrenewable, renewability alone should not be viewed in isolation, since fabrics derived from bamboo, for example, have high-impact waste emissions both to air and water (Fletcher & Grose, 2012). Petroleum-based synthetics should be discontinued in favor of biopolymers, such as corn-based polyester (Fletcher & Grose, 2012), or, alternatively, organic cotton in place of nonorganic, but only to the extent that those materials do not, themselves, have deleterious consequences for the environment. To divert textiles from landfill, fabric should be made from completely biodegradable fibers that are also safe to compost. Moreover, the development of biodegradable fabrics also needs to take into account the production of different waste streams (Fletcher & Grose, 2012) because cross-contamination adversely affects closed-loop systems in which one producer's waste becomes another producer's energy source.

Of course, there is a risk of focusing on new technology and business practices to the exclusion of any fundamental change in our comportment toward the earth. Such tendencies are especially attractive to moderates precisely because they "do not threaten the social relations on which the capitalist market are predicated" (Brooks et al., 2017) and are premised on a concept of "good Anthropocene," the utopian notion that our current global crisis provides "opportunities for new commerce, innovation, and a flourishing of humanity" or the "Let's not let a good crisis go to waste" approach (Brooks et al., 2017). The problem with this view is its faith in a godlike ability of humans to geo-engineer our way out of this predicament (Brooks et al., 2017). Hence, "proposed solutions such as closed loop recycling follow the same flawed logic and do not acknowledge . . . the relentless profit logic, as well as the flawed epistemology of conventional geoengineering approaches" to purportedly mastering the world's natural processes. In contrast, a biomimetic approach to production would be one in which the waste from one plant or industry becomes the resource for another, throughput is continuously cycled, with zero emissions to the environment, and energy is put into maintaining a steady-state of resources and optimizing symbiotic relationships, rather than into rampant growth (Fletcher & Grose, 2012). This means economic growth not just in quantitative terms, but in qualitative terms (Fletcher & Grose, 2012).

However, while the response from industry is encouraging, fundamental change cannot occur in isolation without widespread political support. A women's shoe, comprising recycled materials and designed by activist-outdoor clothier Patagonia to be disassembled and recycled at the end of its life, fell short of its sustainability goal due to the lack of infrastructure that would support such closed-loop processes (Fletcher & Grose, 2012). There needs to be the political will to coordinate public and private initiatives to develop an infrastructure capable of supporting drastic waste reduction, as well as research into what methods and processes actually work, and a means for measuring outcomes and monitoring impacts. At the same time, there also needs to be a multipronged, mutually reinforcing set of culturally sensitive norms that will continue to inspire and guide us along the long road of moving beyond the current destructive consumer model.

For example, in attempting to restore some gravitas to clothing practices, some designers have been attempting to evoke a deeper, more empathic response from potential buyers thereby encouraging the likelihood they will keep their lovingly chosen clothing for years. Some do this by showing their pieces to individual prospective buyers so that they "can take time to savor the unique qualities of each piece and absorb the whole philosophy of the designer in her space" (Fletcher and Grose, 2012). Others involve customers in the design process before a limited run of production. Still others create affordable spaces where customers can have their existing garments repaired or redesigned or create fashionable patterns that customers can download and assemble on their own thereby transforming them from passive consumers into active creators who are more emotionally invested in what they wear. However, empathy involves reflection and "acquired narratives, which build slowly over time" and thus beyond a designer's direct influence. In other words, designers interested in revolutionizing their clients' relationship to their clothing, need reinforcing meta-narratives to help sustain a sense of individual efficacy and involvement in larger collective efforts at sustainability.

FASHIONING A NEW NARRATIVE FOR BIOMIMETIC POLITICS

While it is true that the humanist/environmentalist divide both reinforces "outmoded views of human separateness" from nature, and hobbles creative responses to public policy (Fletcher and Grose, 2012), there are ways to marry what appear to be two sides of a fundamentally opposed dualism consisting of humanism, on the one hand, and environmentalism on the other, so as to muster the collective will to deal effectively with the environmental issues facing us. We clearly cannot assume that exposure to the right information guarantees that consumers will make good decisions and support the best policies. Conservative skepticism about climate change appears to be partially attributable to a conflict between specific ideological values and the most popularly discussed solutions (Campbell & Kay, 2014). Such skepticism toward data that doesn't fit one's ideological values isn't limited to one side of the political spectrum (Campbell & Kay, 2014). Moreover, there is evidence suggesting that those who report being the most concerned about climate change are least likely to engage in pro-environmental behaviors, such as recycling, and the use of public transportation, eco-friendly products, and reusable shopping bags while those who characterized themselves as climate skeptics were, paradoxically, more likely to engage in individual-level environmental actions (Hall et al., 2018). One possible reason for this disparity may be that "highly concerned" participants may engage in moral licensing "whereby their concern about climate change psychologically liberated them from engaging in pro-environmental behavior" (Hall et al., 2018). Hence, although education is important, there needs to be a larger culture—or, more precisely—many overlapping political cultures, to galvanize people and support the wide variety of partial solutions that will be necessary to change our consumerist ways and lessen our environmental impact. Although, in practice, we already have this, at a time where climate change is more divisive than abortion (Leiserowitz et al., 2019), we need to consciously apply biomimetic principles to partisan politics in order to cultivate mutually sustaining relationships "across all aisles" so that we might join forces.

While most dominant political ideologies have had terrible track records when it comes to the environment, there is one thing that they do well and that is the creation of narratives that guide both personal and state action. Up to this point, most dominant political narratives have tended, as Matthew Ally has noted, to increase divisiveness through their perpetuation of false dualities between humanism and naturalism, world and earth, humanity and nature. However, hybrid ideologies may nevertheless come to the same conclusions about the urgency of our environmental situation even if they do so for different reasons. Such hybrid ideologies could be based on the recognition of humankind's need for the planet not, in a technologically driven way of revealing the earth as raw material for the sake of whatever economic system, but rather in a more holistic way stemming from the inherently symbiotic relationship of all living entities, including human beings, within the larger biosphere. We need the earth for sustenance, both materially and existentially. However, according to Ally, the character of this need is not a state but rather a relation that binds organism and environment together. It is a lived relationship that is circular, rather than linear, dialectical, rather than mechanical and one in which each material gesture is part of a larger gestural whole (Ally, 2017). This same circular dialectical need must inform our political culture if we are to search for new distinct, yet mutually intelligible, ways of combining our efforts to transform our relationship to the planet.

This is not as unreasonable an expectation as it might appear, even given our current political climate. There is scholarship that supports the notion that people think, perceive, imagine, and carry out moral decisions using narrative structures (Shenhav, 2006). Indeed, political ideologies embody a narrative view of history and current events. According to Shaul Shenhav (2006), when dealing with two political camps that espouse conflicting, even antagonistic, narratives about what constitutes political reality, what is important is not so much any attempts to change the thematic elements within one side's collective narrative, but rather whether both sides believe that *neither* of their narratives rightly and solely reflects history or "reality" (Shenhav, 2006). The fact that they evince an agreement (to reject the notion that either side has a monopoly on truth), even while leaving the thematic elements of their own narratives intact is important because "such agreement[s] can foster an atmosphere in which each side will feel the need to increase the number of elements of persuasion in its narrative that can affect the other side, rather than staying within the borders of its own . . . 'truth'" (Shenhav, 2006). Such was the insight drawn by Shenhav, after examining an exchange between an Israeli parliamentarian and his Arab political opponent.

Conservative philosopher Roger Scruton has argued that the narrative of the modern environmental movement has largely been controlled by liberals and the Left (Scruton, 2012). This makes sense. Here in the United States, it is out of the 1960s counterculture that Earth Day emerged revitalizing American environmentalism. However, concern for the environment need not be restricted to those left of center. Indeed, it is imperative that it is not. There have long been strands of conservatism that value protection of the environment, despite the Republican Party's skepticism about climate change.[3] However, a narrative focused, for example, on large-scale government intervention and appeals to self-sacrifice are unlikely to move many conservatives who might otherwise be sympathetic to environmentalism.

Although conservatives are at least as likely to support industry-based initiatives as neoliberals (Backer, 2019), which are likely to run counter to the more transformative objectives of deep ecologists, narratives concerning state and local initiatives, developing new technologies that provide jobs, preserving traditional crafts and ways of life meaningful to their constituency, and otherwise developing a narrative based on what Scruton (2012) refers to as *oikophilia*, that is, the love and feeling for home are likely to resonate with them even if it encourages support for certain approaches to the environment that are likely to try the patience of those on the Left. This is not to say that we should abandon the goal of fundamental transformation. It does, however, mean that we need to recognize that successful campaigns to alter our behavior will likely be those that combine appeals to both other-regarding and self-regarding virtues, which may be different according to one's political views (Barnett et al., 2005). Research indicates that people are likely to be moved to change their consumptive behavior based on all sorts of different considerations and it behooves us to gain a deeper understanding of those processes (Barnett et al., 2005). Such incrementalism is admittedly maddening given that many feel that we are running out of time. And yet, what alternative is there?

Given the extent to which technological thinking permeates political ideology, our tendency to blame one or another for the present state of affairs is unproductive. As Scruton points out, capitalism is hardly the only economic system responsible for our current predicament. Some of the worst environmental disasters in the world have been perpetrated under Communist regimes. Like a true technologist, Mao Zedong believed that nature needed to be met with "strategic contempt" (Shapiro, 2001). The results were predictable: Mao's attempt to simultaneously collectivize farming and rapidly increase steel production with the Great Leap Forward unleashed mass deforestation, the destruction of arable land, and a concerted national effort to slaughter all insect-eating sparrows, resulting in the greatest mass famine in world history (Shapiro, 2001). Similarly, the Soviets under Stalin viewed nature "as an enemy to be conquered" in their quest to create a "totally man-made socialist environment" (Shapiro, 2001) which resulted in the destruction of the Aral Sea, widely regarded as one of the worst man-made ecological disasters in history, in a wrong-headed attempt to boost cotton production (Whish-Wilson, 2002).

Unfortunately, there is also a risk in overemphasizing environmental policies rooted in Scruton's *oikophilia.* A deep strand of anti-humanism combining "reactionary ecological themes with anti-immigrant sentiment, eugenic policies, and a nationally or racially-tinged defense of the land" continues to have political traction with certain environmentalists and indicates the extent which all conceptions of nature are socially produced (Biehl & Staudenmaier, 2011). Neither are ideological moderates immune: if Biehl's and Staudenmaier's warnings about the usefulness of the concept of Anthropocene for purposes of effecting a fundamental change in how we comport ourselves to environmental threat, it is because it is a descendant of the same technological thinking that forces the world to appear as raw material for our projects.

Unfortunately, despite the usefulness of Heidegger's philosophy to ecological thinking, his own attempt to harness National Socialism to counter technological thinking was disastrous. His example should serve as a reminder of the importance of maintaining an equilibrium between the most life-affirming aspects of environmentalism and humanism, while jettisoning the worst, most misanthropic and anti-ecological aspects of each respectively. Whether we are able to escape the "desert of empty choice" (Fandozzi, 1982) by developing a new "horizon of significance" against which to relate to the world, while also avoiding the worst aspects of technological thinking remains to be seen but it is certain that a lot will depend upon the stories we tell ourselves and each other about what it means to be a human being, a part of nature, and adorned in dreams.

NOTES

1. Although there is evidence within the fossil record that *Homo neanderthalensis*, a contemporary of Anatomically Modern Human, also engaged in adornment, though not clothing behavior, of more rudimentary nature. See, Juan Luis Arsuaga's *The Neanderthal's Necklace: In Search of the First Thinkers* (1997).
2. Visit: https://www.oed.com/.
3. See the "America's Natural Resources: Agriculture, Energy, and the Environment" segment of the Republican Party's Platform.

REFERENCES

Ally, M. C. (2017). *Ecology and existence: Bringing Sartre to the water's edge.* Lexington Books.

Appleby, J. (2010). *The relentless revolution: A history of capitalism.* W.W. Norton & Company.

Aronoff, M. J. (2001). Political culture. In N. J. Smelser & P. B. Baltes (Eds.) *The International Encyclopedia of the Social and Behavioral Sciences* (Vol. 17, pp. 11640–11644). Elsevier.

Backer, B. (2019, Spring). Conservatives want a voice on the environment. *The Catalyst*, Issue 14. https://www.bushcenter.org/catalyst/environment/backer-young-conservatives.html.

Barnett, C., Cafaro, P., & Newholm, T. (2005). Philosophy and ethical consumption. In R. Harrison, T. Newholm, & D. Shaw (Eds.) *The Ethical Consumer* (pp. 11–24). Sage.

Biehl, J., & Staudenmaier, P. (2011). *Ecofascism revisited: Lessons from the German experience*, second edition. New Compass Press.

Brooks, A., Fletcher, K., Francis, R. A., Dulcie Rigby, E., & Roberts, T. (2017). Fashion, sustainability, and the anthropocene. *Utopian Societies, 28*(3), 482–504.

Calvin, W. H. (2004). *A brief history of the mind.* Oxford University Press.

Campbell, T. H., & Kay, A. C. (2014). Solution aversion: On the relation between ideology and motivated disbelief. *Journal of Personality and Social Psychology, 107*(5), 809–824. https://doi.org/10.1037/a0037963.

Cixous, H. (1994). Sonia Rykiel in translation. In S. Benstock & S. Ferris (Eds.), Deborah Jenson (Trans.) *On fashion* (pp. 95–108). Rutgers University Press.

Claudio, L. (2007, September). Waste couture: Environmental impact of the clothing industry. *Environmental Health Perspectives, 115*(9), A448–A454.

Cline, E. L. (2012). *Over-Dressed: The shockingly high cost of cheap fashion.* Penguin.

Comen, E., & Sauter, M. B. (2019, April 5). The size of a home the year you were born. *24/7 Wall.* https://247wallst.com/special-report/2019/04/05/the-size-of-a-home-the-year-you-were-born-5/11/.

Cook, J., Oreskes, N., Doran, P. T., et al. (2016, April 13). Consensus on consensus: A synthesis of consensus estimates on human-caused global warming. *Environmental Research Letters, 11*(4). doi:10.1088/1748-9326/11/4/048002.

Demeo, S. (1996, January). Dacron polyester: The fall from grace of a miracle fabric. *Science As Culture, 5*(2), 352–372.

Desilver, D. (2018, August 7). For most U.S. workers, real wages have barely budged in decades. *Pew Research Center Fact Tank.* https://pewrsr.ch/2nkN3Tm.

Environmental Protection Agency. (n.d.). Facts and figures about materials, waste, and recycling, nondurable goods: Product-specific data. https://www.epa.gov/facts-and-figures-about-materials-waste-and-recycling/nondurable-goods-product-specific-data#ClothingandFootwear.

Fandozzi, P. R. (1982). *Nihilism and technology.* University Press of America.

Ferrier, M. (2018, July 20). Why does Burberry destroy its products and how is it justified? *The Guardian* (UK). https://www.theguardian.com/fashion/2018/jul/20/why-does-burberry-destroy-its-products-q-and-a.

Fletcher, K., & Grose, L. (2012). *Fashion sustainability: Design for change.* Laurence King.

Food & Agricultural Organization of the United Nations & International Cotton Advisory Committee. (2015). *Measuring Sustainability in Cotton Farming Systems: Towards a Guidance Framework.* http://www.fao.org/publications/card/en/c/c2658c57-5edd-4bd2-bc0d-024ccd8a9785/.

Frankel, J. (2018, May 17). Crisis on the high plains: The loss of America's largest aquifer—Ogallala. *University of Denver Water Law Review.* http://duwaterlawreview.com/crisis-on-the-high-plains-the-loss-of-americas-largest-aquifer-the-ogallala/.

Freeden, M. (1996). *Ideologies and political theory: A conceptual approach.* Oxford University Press.

Freeden, M. (2003). *Ideology: A very short introduction.* Oxford University Press.

FreshwaterCrisis.(n.d.).*InNationalGeographic.* https://www.nationalgeographic.com/environment/freshwater/freshwater-crisis/#close.

Gotmar, V., Singh, P., & Tule, B. N. (n.d.). Wild and cultivated species of cotton. Central In stitute for Cotton Research Nagpur. https://tripleis.org/wp-content/uploads/2019/12/Species-of-Cotton.pdf.

Hall, M. P., Lewis Jr., N. A., & Ellsworth, P. C. (2018, April). Believing in climate change, but not behaving sustainably: Evidence from a one-year long longitudinal study. *Journal of Environmental Psychology, 56*, 55–62. https://doi.org/10.1016/j.jenvp.2018.03.001.

Heidegger, M. (1967). *Being and time.* Trans. J. Macquarrie & E. Robinson. HarperCollins.

Heidegger, M. (1993a). The origin of the work of art. In D. F. Krell (Ed.), A. Hofstadter (Trans.) *Basic writings.* HarperCollins (pp. 139–212).

Heidegger, M. (1993b). The question concerning technology. In D. F. Krell (Ed.), A. Hofstadter (Trans.) *Basic writings*. HarperCollins (pp. 307–341).

Hernandez, E., Nowack, B., Denise, M., & Mitrano, D. M. (2017, May). Polyester textiles as a source of microplastics from households: A mechanistic study to understand microfiber release during washing. *Environmental Science and Technology, 51*(12), 7036–7046.

Interchurch Organisation for Development Cooperation, State Secretariat for Economic Affairs, Textile Exchange, & Helvetas Swiss Interco-operation. (n.d.). The risks of cotton farming. https://organiccotton.org/oc/Cotton-general/Impact-of-cotton/Risk-of-cotton-farming.php.

Jenna, O. (2017, April 8). Looking good can be extremely bad for the planet. *The Economist*. https://www.economist.com/business/2017/04/08/looking-good-can-be-extremely-bad-for-the-planet.

Klein, J. (2016, Winter). Francis Bacon. In E. N. Zalta (Ed.) *The Stanford encyclopedia of philosophy*. https://plato.stanford.edu/archives/win2016/entries/francis-bacon/.

Leiserowitz, A., Maibach, E., Rosenthal, S., Kotcher, J., Bergquist, P., Ballew, M., Goldberg, M., & Gustafson, A. (2019, June 27). Climate change in the American mind: April 2019. Yale Program on Climate Change Communication/George Mason University Center for Climate Change Communication. https://climatecommunication.yale.edu/publications/climate-change-in-the-american-mind-april-2019/.

Magnusdottir, A. (2020, May 13). How fashion manufacturing will change after coronavirus? *Forbes*. https://www.forbes.com/sites/aslaugmagnusdottir/2020/05/13/fashions-next-normal/#2a804c0d78f3.

Moulherat, C., Tengberg, M., Haquet, J. F., & Mille, B. (2002, December). First evidence of cotton at Neolithic Mehrgarh, Pakistan: Analysis of mineralized fibres from a copper bead. *Journal of Archaeology, 29*(12), 1393–1401.

National Cotton Council. (2013, October). Frequently asked questions. http://www.cotton.org/edu/faq/index.cfm.

Postman, N. (1992). *Technopoly: The surrender of culture to technology*. Alfred A. Knopf.

Pulse of the Fashion Industry, 2018. (n.d.). Global Fashion Agenda. http://www.globalfashionagenda.com/download/3700/.

Remy, N., Speelman, E., & Swartz, S. (2016, October). Style that's sustainable: A new fast-fashion formula. *McKinsey & Co*. https://www.mckinsey.com/business-functions/sustainability/our-insights/style-thats-sustainable-a-new-fast-fashion-formula#.

Scruton, S. (2012). *How to think seriously about the planet*. Oxford University Press.

Shapiro, J. (2001). *Mao's war against nature: Politics and the environment in revolutionary China*. Cambridge University Press.

Shenhav, S. R. (2006). Political narratives and political reality. *International Political Science Review, 27*(3), 245–262.

Shim, J., Mangat, P. K., & Angeles-Shim, R. B. (2018, May). Natural variation in wild *Gossypium* species as a tool to broaden the genetic base of cultivated cotton. *HSOA Journal of Plant Science: Current Research*. doi:10.24966/PSCR-3743/100005.

Taylor, C. (1991). *The ethics of authenticity*. Harvard University Press.

United Nations Environment Programme. (2018, November 12). Putting the brakes on fast fashion. https://www.unenvironment.org/news-and-stories/story/putting-brakes-fast-fashion

U.S.DepartmentofAgriculture.(2019,February\22).*CottonOutlook.* https://www.usda.gov/oce/forum/2019/outlooks/Cotton.pdf.

Whish-Wilson, P. (2002). The Aral Sea environmental health crisis. *Journal of Rural and Remote Environmental Health, 1*(2), 29–34.

Wicker, A. (2016, September 1). Fast fashion is creating an environmental crisis. *Newsweek.* https://www.newsweek.com/2016/09/09/old-clothes-fashion-waste-crisis-494824.html.

Zimmerman, M. E. (1990). *Heidegger's Confrontation with Modernity: Technology, Politics, Art.* Indiana University Press.

Zimmerman, M. E. (1993, January). Rethinking the Heidegger-deep ecology relationship. *Environmental Ethics, 15*(3), 195–224.

Part II

POLITICS

Chapter 4

Muslim Perspectives and the Politics of Climate Change

Jennifer Epley Sanders

INTRODUCTION

Religious environmentalism as an ideology and practice is not new as the major religions of the world have historically included some form of concern and care for the environment in their values, principles, sacred texts, rituals, and other customs. Those religious worldviews and traditions remain influential for modern-day environmental discourses, advocacy, movements, and laws. What is arguably different from previous centuries, however, is the large scale and severity of environmental problems and the urgent need for solutions. The current ongoing global environmental crisis involves climate change, deforestation, desertification, toxic waste, biodiversity loss, unsustainable consumption, degraded air/water/land quality, genetic engineering, and ozone depletion. As human beings experience the negative consequences of this extensive environmental crisis, many religious scholars, adherents, and organizations are (re)evaluating their religions' beliefs about and responses to the crisis. Religions have long grappled with how humans should understand and relate to "nature," but now, as Gottlieb (2009) explains, "The sheer scope of this crisis means that *nature*—however it was thought of before this time—has been transformed into something new: the *environment*, that is, a nonhuman world whose life and death, current shape and future prospects, are in large measure determined by human beings." Gottlieb (2009) continues, "Human beings and the environment now form a dialectical totality, each side affecting, and being affected by, the other. If we still depend on nature for food and water, air and minerals, every wild ecosystem depends on some political arrangement for protection, and every living thing is affected by human-made climate change, importation of exotic species, habitat loss, and pollution."

In an effort to improve awareness of the environmental crisis and facilitate faster remedies, interfaith organizations and religious groups are increasingly engaged in environmental education and activism. San Francisco-based Interfaith Power & Light (2017) advocates for a religious response to climate change through education programs and policy advocacy. It has 40 state affiliates in the United States and has worked with over 20,000 congregations on energy conservation, energy efficiency, and renewable energy. New Jersey-based GreenFaith (n.d.) is an international, interfaith environmental nongovernmental organization that helps organize, coordinate, and empower religious and spiritual communities to participate in educational and political campaigns and mobilizations. GreenFaith has partnered with numerous religious and nonreligious groups on international multi-faith sustainable living initiatives, global campaigns dedicated to clean, affordable, and reliable energy, and climate-related mobilizations like the 2018 "Rise for Climate, Jobs, and Justice" in which there were more than 850 events in 95 countries. Religious leaders with large followings have likewise joined the chorus of political voices and participation. Pope Francis's (2015) encyclical on climate change begins with "Saint Francis of Assisi reminds us that our common home is like a sister with whom we share our life and a beautiful mother who opens her arms to embrace us. . . . This sister now cries out to us because of the harm we have inflicted on her by our irresponsible use and abuse of the goods with which God has endowed her." He states, "I urgently appeal, then, for a new dialogue about how we are shaping the future of our planet. We need a conversation which includes everyone, since the environmental challenge we are undergoing, and its human roots, concern and affect us all" (Pope Francis, 2015). The pope's encyclical outlines the types and sources of rapid change and degradation, the social-economic-political dynamics related to ecological problems and problem-solving, and the ways in which Christian principles and teachings can inform educational awareness and new lifestyles. These interfaith and Christian examples of environmental education and activism are not exclusive to a handful of groups, geographic areas, or political contexts. Buddhists, Hindus, indigenous peoples/cultures, Jews, and Muslims from diverse countries and government regimes are tackling the environmental crisis individually and collectively, too.

Although religious environmentalism has been around for a long time, it does not get a lot of national or international media coverage and is still understudied in scholarly social science literature. When religions do get attention, Christianity is often the focus and Islam tends to be overlooked, especially in North American and European settings. Without targeted searches about Islamic environmentalism, it is easy to miss the work of groups like Green Muslims (n.d.), a volunteer-driven nonprofit organization headquartered in Washington, D.C. that connects Muslims to nature and environmental activism. Green Muslims hosts a variety of educational, service, and outdoor recreational events along with connecting the Muslim community to local climate action organizations. This is also why the last decade has seen the development of initiatives like EcoMENA's "Knowledge Bank." EcoMENA (n.d.) is an advisory, consulting, awareness-raising, digital marketing, and publishing organization whose core focus is on the environment and sustainability sectors in the Middle East and North Africa regions. Based in Doha, Qatar, EcoMENA aims to "empower masses with quality environmental information in English as well as Arabic which can be easily accessed by students, researchers, industry professionals, NGOs, project developers, policy-makers, public sector, private sector and most importantly, the general public" through their free online portal that shares articles, reports, analyses, and campaigns. Likewise, Canada-based Khaleafa (n.d.) distributes environmental information, resources, and opinion pieces on its website. Khaleafa specifically educates and mobilizes Muslims through public preaching/sermons via its online Green Khutbah Campaign whose climate plan is "Curbing Consumption, Conserving Energy & Commuting Smarter."

Limited social science research on how Muslims understand the intersection of religion, science, politics, and the environment and why they advocate for Islam as a model for environmental stewardship has negative implications for Muslim communities who want to expand their environmental efforts and for non-Muslims seeking to support those efforts and build cross-cultural collaborations. Thus, the primary aim of this study is to introduce Muslim perspectives about the environment and the politics of climate change to audiences in need of context and data about the religiopolitical actors involved, religious rhetoric, and policy implications. The main sections of this study report on contemporary Muslim public opinion and political participation related to the environment broadly and then in detail on two public religious statements that were released internationally—the "Muslim Seven Year Action Plan on Climate Change" in 2009 and the "Islamic Declaration on Global Climate Change" in 2015.

RELIGIOPOLITICAL ACTORS AND ISLAMIC ENVIRONMENTALISM

As Epley (2010) explains, the concept of "politics" is commonly understood as the "public" sphere in which people engage in a series of power plays over resources, while the concept of "religion" typically connotes a sociocultural system that deals with the "sacred" or "divine" and is "private" in nature. Even though "religion" and "politics" may be understood or treated as distinct concepts, they overlap because "both are concerned with the pursuit of values—personal, social, or transcendent" (Reichley, 1985) and they "are dimensions of human experience engaged in the meaningful exercise of power" (Chidester, 1988). For Muslims, this overlap is real and complete.

Islam is essentially a holistic religion that encompasses all spheres of life including politics (Epley, 2010). Noer (1973) clarifies this union: "Islam, from its inception, has comprised both a religious and a civil and political society. It does not separate the spiritual and the worldly affairs of man, but includes teachings on secular as well as religious activities. Islamic law, *sjari'at*, governs both aspects of life—man's relations with God and his relations with his fellows." In practice, this holistic orientation is evident in the myriad of ways that the religion infuses the daily life of Muslims in ideas, language, imagery, objects, and behaviors. From a Muslim's perspective, Islam frames or is embedded in social, political, and economic life at the individual level, domestic level, and the foreign policy level, particularly in Muslim-majority countries. As such, Muslims are religiopolitical actors where "religiopolitical actors" refers to religious persons and groups who operate in both the "religion" and "politics" spheres or view and experience the two realms as one (Epley, 2010).

When it comes to Islamic environmentalism (also known as "Green Islam" or "Eco-Islam"), religiopolitical actors do not always have rigid boundaries separating "religious" issues from "political" issues. The environmental crisis involves everyone *and* threatens everyone practically and theologically. Personal, household, and community concerns include, for example, the "earthly" matters of health, education, and jobs as well as the "spiritual" matters of one's own connection and responsibilities to God and a community's obligations toward God and each other. These matters are combined with a sense of individual agency alongside an acceptance and encouragement of collective action due to Islam's emphasis on both personal and social welfare.

While Islam as a religion has universal characteristics and is unifying, Muslims themselves are diverse with multidimensional identities. The global Muslim community consists of a wide range of cultural, ethnic, racial, and national groups. Differences based on sects, socioeconomic status, education, political affiliations, age, gender, and other groupings contribute to the heterogeneity. Brown (2000) adds, "Islam knows no 'church' in the sense of a corporate body whose leadership is clearly defined, hierarchical, and distinct from the state." There is no single institutionalized authority commanding conformity. This intra-religious diversity and decentralization contribute to variation in the forms, functions, and outcomes of Islamic environmentalism worldwide. Individuals, households, organizations, communities, and governments determine for themselves by and large the best courses of action to combat climate change and other environmental problems. This means that Islam's greater community recognizes that public opinion and political participation connected to the environment are (and should be) adapted based on different needs, cultures, resources, government structures, and time periods. At the same time, intra-religious diversity and decentralization arguably limit efforts to create a cohesive environmental movement within a country or among countries.

MUSLIM PUBLIC OPINION AND POLITICAL PARTICIPATION

According to the Pew Research Center (2017a), there were an estimated 1.6 billion Muslims in the world as of 2010. This makes Islam the second largest religion after Christianity. Around 62 percent of Muslims overall live in the Asia-Pacific region, while the highest concentration of Muslims lives in the Middle East-North Africa region. About 93 percent of the latter's population are Muslim compared to 30 percent in sub-Saharan Africa and 24 percent in the Asia-Pacific region (Pew Research Center, 2017a). The Pew Research Center identifies fifty Muslim-majority countries and territories using the standard of 50 percent or more of a population self-identifying as Muslim. The top four countries with the largest number of Muslims are Indonesia, Pakistan, India, and Bangladesh (Pew Research Center, 2017b).

While Muslims constitute a sizable share of the world's population, public opinion and political participation data for their demographic are difficult to obtain. Comprehensive survey data from all fifty Muslim-majority countries and territories are unavailable or incomplete due to infrequent international and national surveys conducted by scholars and governments, limited physical and personnel resources, higher demand for single-country studies than comparative country studies, preferences for short-term election-related surveys, and restricted access such as data collected and reserved for private political or commercial purposes. Using the World Values Survey is one way to try to fill part of this knowledge gap. The World Values Survey conducted its Wave 6 questionnaire (2010–2014) in sixty countries and territories, which included fourteen Muslim-majority countries: Algeria, Azerbaijan, Iraq, Jordan, Kyrgyzstan, Lebanon, Malaysia, Morocco, Nigeria, Pakistan, Tunisia, Turkey, Uzbekistan, and Yemen. For each of these countries, there were between 600 and 1,600 respondents who self-identified as Muslim in each survey's random sample of a population (Inglehart et al., 2014).

Although the World Values Survey contains valuable data, there are two caveats for its use. First, the questionnaires do not usually provide all respondents with standardized definitions, which can leave question wording and answer choices open to interpretation by respondents and researchers. Second, the questionnaires do not ask supplemental open-ended questions about a respondent's answers to the closed-ended questions. In other words, respondents do not thoroughly describe the development of or context for their beliefs and behaviors. The World Values Survey codebook does not give rationales for its survey design process and the construction of its question and answer items, so possible reasons for primarily utilizing closed-ended questions and truncating answer choices include, but are not limited to, time constraints, questionnaire length, anticipated respondent fatigue, and potential cultural or political sensitivities. Despite these two caveats, the World Values Survey is a rich dataset and suitable as a starting point for Muslims' perspectives and participation related to religion and the environment.

Variable 25 in the World Values Survey asks respondents about their membership in religious organizations. Variable 30 asks them about membership in environmental organizations. The two variables have three answer choices: Active member, Inactive member, or Not a member. For each of the fourteen Muslim-majority countries surveyed, respondents' active membership in religious organizations ranges from a low of 0.6 percent up to 16 percent with the exception of Nigeria which lists 74.6 percent as "active members." The percentage range for active membership in environmental organizations is 0.2–6.9 percent. Most of the Muslim respondents, with the exception of Nigerians, are not active members, but might still feel some kind of affiliation since inactive memberships range from 1 percent to 15.8 percent for religious organizations and span 0.1–12.4 percent for environmental organizations.

Low percentages of active memberships in religious and/or environmental organizations are not unusual. In 2018, the Pew Research Center conducted their Global Attitudes Survey in fourteen countries: Argentina, Brazil, Greece, Hungary, Indonesia, Israel, Italy, Kenya, Mexico, Nigeria, Philippines, Poland, South Africa, and Tunisia. According to that survey, most people have voted at least once in the past (e.g., median of 78 percent), but other forms of participation such as attending political campaigns or speeches, participating in volunteer organizations, posting comments about political issues online, protesting, or donating money to social or political organizations are much less common (Wike & Castillo, 2018). Although the Pew Research Center finds that respondents can be motivated to participate on issues like health care, poverty, and education, much still depends on variables such as age, education, online social networking, and placement on a left-right ideological scale (Wike & Castillo, 2018).

Besides asking about involvement in voluntary organizations, the World Values Survey requests that respondents share their environmental perspectives. The results for three different questions follow below.

Variable 78 provides a description—"Looking after the environment is important to this person; to care for nature and save life resources"—and asks respondents to indicate whether that person is very much like you, somewhat like you, not like you, or not at all like you. For this self-assessment, Muslim respondents in Jorden and Uzbekistan have the highest percentages with over 45 percent choosing "Very much like me." Variable 80 asks respondents to select from a set of problems that they consider to be the most serious one for the world as a whole. The answer choices are "People living in poverty and need," "Discrimination against girls and women," "Poor sanitation and infectious diseases," "Inadequate education," or "Environmental pollution." Although there are respondents who think that environmental pollution is a serious problem for the world, the majority of respondents in most of the countries selected "People living in poverty and need." The World Values Survey also asks respondents about their confidence in a variety of organizations such as the armed forces, the press, television, police, courts, political parties, universities, major companies, charitable or humanitarian organizations, and the United Nations. Variable 122 is environmental organizations, and as table 4.1 shows, a considerable percentage of those surveyed selected "A great deal" or "Quite a lot" instead of "Not very much" or "None at all." Muslim respondents in Malaysia, Pakistan, and Uzbekistan appear to have higher confidence in environmental organizations than their Muslim-majority country counterparts.

In an attempt to better understand respondents' everyday priorities, Variable 81 of the World Values Survey asks respondents to pick one of two statements that are closer to their own point of view regarding the environment and economic growth and jobs, which table 4.2 illustrates.

Kyrgyzstan, Malaysia, Turkey, and Uzbekistan are countries in which a larger percentage of survey respondents chose "Protecting the environment should be given priority, even if it causes slower economic growth and some loss of jobs" than "Economic growth and creating jobs should be the top priority, even if the environment suffers to some extent." Although significant percentages of respondents prioritized the environment for this question item, their perspectives do not coincide exactly with picking "environmental pollution" as the most serious problem for the world in table 1 since "People living in poverty and need" seemed to be the most pressing issue.

As for political participation related to the environment, the World Values Survey asks two questions. Variable 82 inquires about giving money to ecological organizations during the past two years. The percentage of Muslim respondents selecting "Yes" ranges from 1.7 percent to 18.5 percent depending on the country. Variable 83 solicits responses for participating in any demonstrations for environmental causes during the past two years. The percentage range for answering "Yes" to Variable 83 is 2.1–20 percent. Muslim respondents in Lebanon, Malaysia, Nigeria, and Pakistani Muslims in general and Sunni Muslims in specific appear to be somewhat more engaged on these two measures of political behavior compared to respondents in the other

Table 4.1 Muslims' Environmental Perspectives

	Describes person like me: "Looking after the environment is important to this person; to care for nature and save life resources." (%)		*Most serious problem for the world as a whole (%)*	*Confidence in environmental organizations (%)*	
Country	*Very much like me*	*Like me*	*Environmental pollution*	*Great deal*	*Quite a lot*
Algeria	27.6	26.8	11.2	12	23
Azerbaijan	17.7	27.6	8.1	8.3	40
Iraq	36.4	32.2	12.2	15.4	25.8
Jordan	45.7	32.7	3.8	8.8	32.3
Kyrgyzstan	21.4	31.1	12.2	14.1	41.3
Lebanon	25	27.8	7.8	12.5	32.8
Malaysia	25.5	37.1	15.4	18.4	61.1
Morocco	24.3	38.3	7.1	27.2	34.6
Nigeria	34.8	30.5	0.1	18.5	41
Pakistan	15.8	31.6	0	0	57.9
Pakistan—Sunni	18.6	32.9	0.4	6.7	32.7
Tunisia	27.1	19.8	4.1	10.8	17.2
Turkey	24.7	39.7	4.2	13.7	39.2
Uzbekistan	46.1	29.9	22.4	46.5	40.3
Yemen	25.7	41.3	3.6	8.4	23.8

Source: Table created by author using data from Inglehart et al., 2014.

countries. Similar to the aforementioned low percentages of active memberships for environmental organizations, most respondents in the surveyed Muslim-majority countries did not report lots of financial contributions to ecological groups or involvement in demonstrations for environmental causes. One methodological challenge with this data is that comparisons to political participation on other issues of concern are not possible with the World Values Survey because variables 85 through 94 which mention petitions, boycotts, peaceful demonstrations, strikes, and other acts of protest are not tied to issue areas or public policies in their question wording. The survey simply asks if respondents "Have done, "Might do," or "Would never do" those forms of political action and then asked about frequency for any respondents who said that they had done an activity.

As the previous data shows, although Muslims might have an interest in or an awareness of environmental issues, there is a tendency to have low levels of political participation. This is a common pattern around the world

Table 4.2 Muslims' Points of View when Discussing the Environment and Economic Growth

	Discussing environment and economic growth (%)	
Country	*"Protecting the environment should be given priority, even if it causes slower economic growth and some loss of jobs."*	*"Economic growth and creating jobs should be the top priority, even if the environment suffers to some extent."*
Algeria	38.5	52.4
Azerbaijan	31.4	66.3
Iraq	44.2	50.2
Jordan	37.2	62
Kyrgyzstan	63.3	36.7
Lebanon	41.8	57.3
Malaysia	74.8	21.6
Morocco	66	26.3
Nigeria	33.7	64.9
Pakistan	21.1	63.2
Pakistan—Sunni	47.9	51.3
Tunisia	35.7	64.3
Turkey	51.7	48
Uzbekistan	66.8	30.4
Yemen	38.3	61

Source: Table created by author using data from Inglehart et al., 2014.

for religious and nonreligious populations regardless of the topic or policy matter at hand, however. Exceptions include voting in local and national elections or protest behavior during social and political crises. Individual and group calculations for political engagement often involve questions about resources and risks since different types of government systems, laws, social customs, economic costs, majority-minority politics, proximity to environmental problems, and short- or long-term time horizons factor into personal decisions about political education and engagement overall and environmental activism in particular. Even if there are low levels of political participation in a population, this does not necessarily imply a complete disregard for the environment and lack of action on the part of individuals to improve the environment. Recycling, reducing waste, reusing materials, and educating one another are likely occurring in households and schools, for example, but researchers do not yet have sufficient data to draw conclusions about the distribution of such activities and its effectiveness at local, state, national, or international levels. Similarly, there are methodological challenges to collecting and analyzing data for the small group of committed Muslim environmental educators and activists who do engage in politics at multiple levels.

The preceding statistics give us a general introduction to Muslim perspectives and practices about and for the environment. The next section of this study shifts from public opinion data to specific cases of religious rhetoric to gain added insights into the function, forms, and outcomes of Islamic environmentalism. A focus on contemporary messaging and the selected strategies of several Muslim leaders and groups for the politics of climate change is another way to explore how Muslims understand the intersection of religion, science, politics, and the environment and why they advocate for Islam as a model for environmental stewardship.

RELIGIOUS RHETORIC

The Language and Roots of Stewardship

Islamic environmentalism takes different forms from country to country but shares the same religious rhetoric and origins. Contemporary "Green Islam" has its roots in the seventh-century teachings of the Prophet Muhammad. Wihbey (2012) writes: "The major Islamic texts themselves, the Koran (or Quran) and Hadith, are replete with references to nature and its sacredness. Various stories and sayings associated with the Prophet encourage thoughtful stewardship and suggest care for God's creation, and there's a strong intellectual basis for a kind of proto eco-Islam, scholars note." Khattak et al. (2019) emphasize how stewardship means being "viceregents on earth" or "viceroys of God": "One of the fundamental Islamic beliefs is that whatever we own and is given in our possession basically belongs to the Allah Almighty. This belief is further strengthened by another well-known belief that we are trustees of everything given to us by Almighty." It then follows that taking care of the environment is one of the responsibilities expected of Muslims as trustees.

Scholars and adherents point to passages from sacred texts, oral traditions, Islamic law, and historical practicalities as evidence, explanation, and motivation or inspiration for submission and service to God through words and deeds concerning the environment. Dehlvi (2020) notes that there are more than 500 verses in the Holy Qur'an that deal with the natural phenomenon and that Islamic jurisprudence contains regulations concerning the conservation and protection of water resources, land, and wildlife. A sample of passages from different surahs (chapters) and ayahs (verses) from the Holy Qur'an (*The Noble Qur'an*, n.d.) that are associated with environmental ethics and responsibilities include:

- Qur'an 2:60—"And [recall] when Moses prayed for water for his people, so We said, 'Strike with your staff the stone.' And there gushed forth from it twelve springs, and every people knew its watering place. 'Eat and drink from the provision of Allah, and do not commit abuse on the earth, spreading corruption.'"

- Qur'an 6:38—"And there is no creature on [or within] the earth or bird that flies with its wings except [that they are] communities like you. We have not neglected in the Register a thing. Then unto their Lord they will be gathered."
- Qur'an 6:99—"And it is He who sends down rain from the sky, and We produce thereby the growth of all things. We produce from it greenery from which We produce grains arranged in layers. And from the palm trees—of its emerging fruit are clusters hanging low. And [We produce] gardens of grapevines and olives and pomegranates, similar yet varied. Look at [each of] its fruit when it yields and [at] its ripening. Indeed in that are signs for a people who believe."
- Qur'an 6:165—"And it is He who has made you successors upon the earth and has raised some of you above others in degrees [of rank] that He may try you through what He has given you. Indeed, your Lord is swift in penalty; but indeed, He is Forgiving and Merciful."
- Qur'an 16:68-69—"And your Lord inspired to the bee, 'Take for yourself among the mountains, houses, and among the trees and [in] that which they construct. Then eat from all the fruits and follow the ways of your Lord laid down [for you].' There emerges from their bellies a drink, varying in colors, in which there is healing for people. Indeed in that is a sign for a people who give thought."

Moreover, lessons from well-known hadith (traditions) refer to being rewarded for helping living creatures, loving each another and having mercy on those who live on earth, God asking why someone killed a sparrow or any other creature without its deserving it, "grievous things" include the "killing of breathing beings," and how being kind to God's creatures is being kind to oneself (Rahman, 2017). One familiar story from the life of the Prophet recounts how, during a journey, one of Muhammad's companions removed a baby pigeon from a nest. Muhammad confronted the thief and gently returned the bird to its nest. "For charity shown to each creature with a wet heart, there is a reward," the Prophet declared (Smith, 2002).

Historical necessity and geographic context contributed to close ties between religion and the environment, too. "The advent of Islam as an organized religion occurred in the desert environment of Arabia, and hence there was considerable attention paid to ecological concerns within Islamic ethics. There is a reverence of nature that stems from essential pragmatism within the faith," says Dr. Saleem Ali (Gelling, 2009). The language of and rationales for Islamic environmentalism are thus not from the "outside" in terms of other philosophies, ideologies, or sciences. In present-day parlance, Islam has had a "grassroots" orientation and approach to environmental stewardship characterized by inclusiveness, moderation, balance, and charity from the beginning.

While environmentalism is deeply embedded in the foundational principles and guidelines of Islam, there are practical challenges in dealing with the politics of climate change and implementing solutions. For individual Muslims, Yildirim (2016) notes, "The circumstances under which many Muslims live are so severe that it is not at all surprising that the environment takes a back seat, regardless of the relatively high rank it holds in Islam." At the group-level, building pro-environment organizations, schools, businesses, and governments takes a massive amount of sustained interest, resources, coordination, and maintenance over time. Despite these challenges, Muslim groups have attempted to collaborate domestically and internationally to fulfill their stewardship responsibilities.

Muslim Seven Year Action Plan on Climate Change (2009)

The British nonprofit organization Earth-Mates Dialogue Center and the Kuwait Ministry of Awqaf and Islamic Affairs drew up the "Muslim Seven Year Action Plan on Climate Change: 2010–2017" in 2009. It incorporated contributions from participants in a 2008 workshop in Kuwait that had twenty-two participants from Islamic NGOs, academics, government figures, Muslim environmental activists, and environmental specialists from fourteen Muslim countries and communities (Alliance of Religions and Conservation, ARC, and United Nations Development Programme, UNDP, 2009). Religious communities determined their own metrics for each action plan based on suggested ideas from a handbook provided by the ARC and UNDP. According to the Earth-Mates Dialogue Center, around fifty religious scholars from across the Muslim world endorsed the long-term plan. Dr. Youssef Al Qaradawi, the president of the International Union of Muslim Scholars at the time, supported the Muslims who convened in Istanbul, Turkey from July 6–7, 2009. The plan advocated for the establishment of an institutional framework, developing an overall capacity to deal with climate change and environmental conservation, developing and enhancing communication, outreach, and partnerships, as well as activating and reviving implementation of previous initiatives, plans, and declarations. The group's vision concentrated on making a world that is environmentally safe for children and the next generations and creating a world in which all nations and religions can live in harmony with nature and enjoy justice and fairness. The group's mission centered on the Muslim community contributing to ongoing global efforts dealing with climate change in ways that reflect Islamic principles and values.

The proposed "Muslim Seven Year Action Plan" recommended establishing an umbrella organization called the "Muslim Association for Climate Change Action" (MACCA) to be based in London that would manage and implement the plan. The aim was to have MACCA's first meeting by the end of November 2009 in Jakarta, Indonesia with seventy-five representative members from different countries and communities. MACCA was expected to complete the following tasks:

- Create a Wakf in one year in order to implement the Climate Change plan;
- Establish Islamic labels for different products. This would be an Islamic environmental labelling system with strict authenticity standards;
- Work toward a "Green Hajj" with the Saudi minister of the Hajj. Aim to have the Hajj free of plastic bottles after two years and introduce environmentally friendly initiatives over the next five to ten years to transform the Hajj into a recognized environmentally friendly pilgrimage. The vision is that pilgrims on the Hajj will take back an understanding of care of creation as an act of faithfulness;
- Pilot the construction of a "green mosque" to showcase best practice in heating, light, design, and so on. Plan to use this as a model for building other mosques worldwide;
- Develop two to three Muslim cities as "green cities" which can act as a role model for greening other Islamic cities. Select ten cities in the Muslim world to be greened after the success of the first phase;
- Focus on education on the environment:
  - Make more material on the conservation of the environment available to places of Islamic learning, focusing on the training of imams and in schools;
  - Develop guidebooks for teachers in primary, middle, and secondary schools over the next three years;
  - Develop educational materials for nonformal education in the next three years;
  - Prepare guidelines and train imams on environmental conservation and climate change issues;
  - Sponsor ten postgraduate students to work on Islam and climate change over the next five years;
  - Establish a chair for professorship in dealing with climate change.
- Develop a best practice environment guide for businesses;
- Apply environmental principles in the publication of the Quran. Work toward printing a "green Quran" on paper that comes from sustainable wood supplies;
- Re-introduce Islamic rituals from an environmental perspective. Use the Hajj season to distribute these ideas and the Friday Khotbas;

- Establish a special TV channel for Islam and the environment to be broadcast in different languages;
- Develop an international prize for research related to environmental conservation (Earth-Mates Dialogue Center, n.d.).

Around 200 Muslim scholars, experts, and representatives of Islamic civil society organizations, environmental ministry delegates, and religious and charitable foundations from Bahrain, Indonesia, Kuwait, Morocco, Senegal, and Turkey endorsed the climate action plan (Earth-Mates Dialogue Center, n.d.). It is unclear what roles the signatory Muslim scholars would play beyond endorsing the plans, but their support was still important for showing the credibility and validity of the plan.

The plan to establish MACCA was approved at the July 2009 conference. MACCA was then officially launched in November 2009 at the ARC and UNDP's summit titled "Many Heavens, One Earth: Faith Commitments for a Living Planet," which Prince Philip, Duke of Edinburgh, hosted. Membership was open to individuals, NGOs, government agencies, and other interested parties, with a secretariat housed at the Earth-Mates Dialogue Center (Berkley Center for Religion, Peace, and World Affairs, n.d.). It is unclear, however, if MACCA still exists or just existed for a brief period. There are contradictory online posts about results from the International Muslim Conference on Climate Change held in Bogor, Indonesia in 2010. Most sources claim that MACCA did not get formally set up at the Bogor meeting (International Muslim conference on climate change pressures OIC to act, 2010), but other websites in later years give the impression that Muslim groups intermittently reference the work of MACCA either physically or digitally.

The "Muslim Seven Year Action Plan" looked to establish, share, model, and support best practices for combating climate change. It strategically focused on Muslim communities to leverage their shared religious principles, rhetoric, and practices. The plan included tangible objectives, activities, and measures, too. The plan arguably raised more questions and concerns than answers, though. Who would be responsible for the enormous costs of the plan? Would MACCA members be the sole contributors to the *wakf/waqf* (an endowment made by Muslims for religious, educational, and charitable causes)? Which Muslim individuals, associations, or government entities would be called on to donate, how much, and how frequently? Which Islamic rituals would be re-introduced from an environmental perspective? Who would oversee interpreting that process? Would faith, science, or a combination of the two frame and inform those Islamic rituals? How would conflicts be arbitrated and resolved? What resources would be available to bridge the economic, education, and/or gender gaps for participation among different countries? What about the roles of "average citizens" who may or may not reside in Muslim-majority cities and countries?

While Islam is ubiquitous throughout the world, it is not homogenous. It is therefore difficult for any single plan to be implemented worldwide, particularly if one factors in diverse country-level needs, interests, and constraints as well. Although organizations located within the participating countries have met some of the goals associated with the "Muslim Seven Year Action Plan," the process has been much more gradual, small scale, and fragmented than originally anticipated. There are currently no systematic studies that analyze the conditions under which aspects of the plan were likely to be "successful"; to what extent the religiopolitical actors involved were motivated by "self-interest," "environmental ethos," or other reasons; whether or not the plan served as a foundation for later debates, proposals, and declarations like the "Islamic Declaration on Global Climate Change"; or if the plan influenced the eventual development of other alliances such as the Global Muslim Climate Network. Additionally, there is missing data and analysis regarding the impact of closures such as the ARC ending in June 2019 after twenty-three years of connecting projects and programs between the major religions, international organizations, and nongovernmental organizations working on environmental issues (Alliance of Religions and Conservation, 2019).

The "Muslim Seven Year Action Plan" itself was not an historical anomaly because about six years later, another set of faith leaders, international development policy-makers, academics, and other experts attempted their own international declaration based on Islamic environmentalism. Some of the religious ideas and rhetoric overlapped between the two documents, but they diverged in a few substantial ways.

Islamic Declaration on Global Climate Change (2015)

On August 18, 2015, Islamic Relief Worldwide, GreenFaith, and the Islamic Foundation for Ecology and Environmental Sciences co-sponsored the International Climate Change Symposium in Istanbul, Turkey in which Muslim leaders, specialists, and policy-makers adopted the "Islamic Declaration on Global Climate Change." They called on government representatives participating in the Conference of the Parties (COP) meeting in Paris, France for the UN Climate Change Conference in December 2015 to consider: (1) The scientific consensus on climate change to stabilize greenhouse gas concentration in the atmosphere at less dangerous levels, (2) the need to set clear targets and monitoring systems, (3) acknowledging the dire consequences if we do not do so, and (4) the responsibility the COP has on behalf of humanity (United Nations Climate Change, 2015).

The symposium's alliance intentionally called on the world's 1.6 billion Muslims to play an active role in combating climate change (United Nations Climate Change, 2015). The preamble of the declaration incorporates religious rhetoric such as "We human beings are created to serve the Lord of all beings, to work the greatest good we can for all the species, individuals, and generations of God's creatures" (International Islamic Climate Change Symposium, 2015).

This type of statement immediately frames the politics of climate change in two ways: (1) Humans are in service to and at the mercy of God, and (2) everyone must work individually and together toward the common good. This kind of framing also allows space for participation regardless of identity or status and that interfaith efforts and working with scientists (who are religious or secular) are possible.

The writers of the declaration go on to directly identify human beings as the source of the environmental crisis. The document avoids drawing attention to certain "culprits" such as governments or corporations, but later discusses ways in which such institutions can change and benefit the environment. This approach differs from other religious and nonreligious groups who readily and strongly blame named entities right from the outset. This section of the declaration also takes the long view of climate change challenges by asking about the implications for the next generations on earth as well as what happens upon our own passing and meeting God. This joint appeal pulls at Muslims' concerns and feelings about their current family members, friendships, and communities *and* the afterlife. Additionally, environmental degradation in this case is a clear result of not fulfilling one's religious obligations as a caretaker or steward of the earth, which ultimately means that one has not been a good subject of God.

The declaration goes on to mention evidence-based warnings for climate change from the 2005 UN Environment Programme's Millennium Ecosystem Assessment that was supported by over 1,300 scientists from ninety-five countries and the 2014 Intergovernmental Panel on Climate Change comprising of representatives from over a hundred nations. The document stresses that humankind cannot afford the slow progress or deadlocks that have occurred during and since those warnings were published. The declaration further expresses alarm that the successor to the Kyoto Protocol, which should have been in place by 2012, has been delayed and recommends that all countries, especially the more developed nations, increase their climate change efforts (International Islamic Climate Change Symposium, 2015).

The Muslim leaders involved with the declaration emphasize the role of science in understanding climate change, the potential for cooperation between religious groups and the scientific community, and that time is of the essence. They once again place the responsibility on everyone to do their part and note the essential participation of all countries to be proactive instead of simply being reactive or unengaged. For non-Muslim audiences, these remarks oppose existing negative stereotypes of Islam and Muslims as being "anti-science" and "insular."

In sections 2.1–2.8, the declaration points to multiple specific passages in the Qur'an to affirm the following:

2.3 We affirm that—
- God created the earth in perfect equilibrium (m z n);
- By His immense mercy we have been given fertile land, fresh air, clean water and all the good things on Earth that make our lives here viable and delightful;
- The earth functions in natural seasonal rhythms and cycles: a climate in which living beings—including humans—thrive;
- The present climate change catastrophe is a result of the human disruption of this balance;

and

2.6 We recognize that we are but a minuscule part of the divine order, yet within that order we are exceptionally powerful beings, and have the responsibility to establish good and avert evil in every way we can. We also recognize that—
- We are but one of the multitude of living beings with whom we share the earth;
- We have no right to abuse the creation or impair it;
- Intelligence and conscience should lead us, as our faith commands, to treat all things with care and awe (taqw) of their Creator, compassion (rahmah) and utmost good (ihs n). (International Islamic Climate Change Symposium, 2015)

These sections address responsibility and accountability for Muslims and non-Muslims alike, but the Qur'anic evidence is mostly aimed at persuading Muslims to be more motivated and action oriented. In this way, a Muslim may be moved to participate in positive climate change efforts because of the real, tangible consequences currently experienced on earth, as well as, if not more importantly so, because of religious justifications, namely God's expectations and commands given that the Qur'an is the central religious text in Islam.

The third portion of the declaration calls on representatives attending the 2015 United Nations Climate Change Conference; well-off nations and oil-producing states; the people of all nations and their leaders; corporations, finance, and the business sector; and all groups to join in collaboration, cooperation, and friendly competition to take concrete action toward addressing climate change challenges. Sections 3.2–3.4 suggests steps to reduce consumption, where and who to invest in, commitment to renewable energy and zero emissions, models of well-being, and how to heed social and ecological responsibilities (International Islamic Climate Change Symposium, 2015). In section 3.5, the declaration highlights alliances: "We welcome the significant contributions taken by other faiths, as we can all be winners in this race. . . . If we each offer the best of our respective traditions, we may yet see a way through our difficulties" (International Islamic Climate Change Symposium, 2015). Section 3.6 has a special call to all Muslims wherever they may be (e.g., heads of state, political leaders, the business community, religious leaders and scholars, mosque congregations, educators, community leaders, civil society activists, and the media) to "tackle habits, mindsets, and the root causes of climate change, environmental degradation, and the loss of biodiversity in their particular spheres of influence" (International Islamic Climate Change Symposium, 2015).

Together all of these sections provide the overall principles, guidelines, and goals for everyone to aim for and implement regarding climate change. There is a combination of social, economic, and religious reasons and evidence offered that collectively is convincing or individually tailored to sway someone depending on the reader's background and preferences. The interfaith approach is arguably more palatable for non-Muslims, while at the same time, instructive for Muslims to cooperate with diverse communities as the document makes points and incorporates evidence that is agreeable to Muslims.

Overall, the declaration attempts to strike a balance in its eight pages between religious or religiopolitical actors and nonreligious actors along with their various interests and strategies. The document has a unifying and inclusive tone. It focuses on what can be done nowadays to combat climate change rather than on who is mostly at fault in part because we have all contributed to the environmental crisis. Using a unifying and inclusive tone and looking ahead instead of excessively bringing up the past might also possibly be a political maneuver to solicit cooperation instead of garner resistance. The declaration makes use of accessible language to understand the intersection of religion, the environment, and politics and proposes what the authors believe to be manageable and meaningful measures. Critics could dismiss the hopeful, idealistic impressions and intentions of the document and complain about the difficult practicalities of collaboration, the huge scale of change and resources required internationally, the slow pace of a government or country's timelines, or dealing with skeptics who deny the realities and impact of climate change. While the document's religious and political rhetoric may have limited reach, we do not yet have sufficient data to automatically dismiss its potential and actual contributions to individual-level shifts in thinking and action or public policy changes at the domestic and international levels. Future research studies concerning public and/or institutional reactions to the declaration would facilitate tracking what differences, if any, the document made for individuals and groups.

POLICY IMPLICATIONS

The two aforementioned public statements contain different examples and degrees of religious rhetoric around climate change. The documents provide readers with an introduction to how Muslims view prospective solutions to climate change challenges and how they justify the importance of acting and acting quickly. Both statements avoided an "us" versus "them" attitude and instead suggested ways to bring people together within Muslim communities and between Muslims and non-Muslims. This could be a product of Islam's orientation toward the public good and/or different cultures' approaches to collectivism. For representatives and government entities at the local, state, national, and international levels, the documents are potentially useful as a source of Muslim voices for those communities' needs and wants, which are often formally underrepresented or misrepresented. Moreover, the documents may contain religious rhetoric that parallels other religious, spiritual, or moral/ethical traditions, thus reinforcing connections between religion, ethics/morality, science, and the environment and possibly adding to collective action. The statements might be sources of new, creative, and persuasive lines of argumentation and strategies, too. In short, the statements add diverse viewpoints to the existing marketplace of ideas.

What goes unaddressed in both statements, however, are explicit references to or policy recommendations involving differences across important social, economic, and political cleavages. For instance, what would the participants in the previous conferences have to say about ethnicity, gender, socioeconomic status, political orientation, religious interpretations, linguistics, and local history and traditions? These variables may contribute to perceptions of "insider" and "outsider" identities and then internal or external conflicts that inhibit communication and cooperation.

Another concern for the policy-making process is how to get followers and partners to really prioritize climate change agendas and solutions? There are countless competing problems and limited resources facing humanity, especially in certain regions of the world, that it is difficult to imagine a shared, collective awareness and feasible global action plan. This is reminiscent of challenges that international organizations like the UN and the World Health Organization face on a regular basis.

One other consideration is how individuals and governments might go about mediating the differences between relativist, universal, and pragmatic values. Religions and secular institutions often have elements of all three, so which would or should be upheld, when, and why concerning climate change? Interrelated with the values question is thinking about political ideologies or government regime types and the ways in which religion—in this case Islam—"fits" or does not "fit" with such perspectives and institutions for creating and implementing climate change policies.

CONCLUSION

It is difficult to generalize Muslims' perspectives and political participation for environmental politics because of intra-religious diversity, multidimensional identities, country and regional differences, and research limitations. Much more research is necessary to accurately and completely identify and confirm what Muslims believe about how their religion is related to the environment and public policy along with what practices they deem are "acceptable" and "encouraged" at the individual, group, and country levels. There are not a lot of publicly available English-language source materials that report on different types of religiopolitical participation by Muslims for the environment and how widespread their activities are global. Future research projects could expand on this study by using both quantitative and qualitative methods in multiple Muslim-majority countries to see if religious rhetoric and international statements like the "Muslim 7 Year Action Plan on Climate Change" and "Islamic Declaration on Global Climate Change" had any short-term or long-term impacts on individual and group beliefs and/or engagement. Perhaps interdisciplinary teams of researchers could also incorporate the role of institutions, geography, cultural or social resources, economics, and political systems into their analyses to better identify the nuances and limits of how religion and politics interact or overlap when it comes to the environment.

REFERENCES

Alliance of Religions and Conservation. (2019, June 27). ARC has closed after 23 years. http://www.arcworld.org/news.asp?pageID=905.

Alliance of Religions and Conservation and United Nations Development Programme. (2009, November). Many heavens, one earth: Faith commitments to protect the living planet. http://www.arcworld.org/downloads/Faith-Commitments-Handbook.pdf.

Berkley Center for Religion, Peace, and World Affairs. (n.d.). Muslim association for climate change action. https://berkleycenter.georgetown.edu/organizations/muslim-association-for-climate-change-action.

Brown, C. L. (2000). *Religion and state: The Muslim approach to politics*. Columbia University Press.

Chidester, D. (1988). *Patterns of power: Religion and politics in American culture*. Prentice Hall. Quoted in M. Corbett & J. Mitchell Corbett. (1999). *Politics and Religion in the United States*. Garland Publishing, Inc.

Dehlvi, S. (2020, April 3). Islam—in harmony with nature. *EcoMENA*. https://www.ecomena.org/islam-nature/.

Earth-Mates Dialogue Center. (2009). Istanbul declaration of the Muslim 7-year action plan on climate change: 2010–2017. http://www.arcworld.org/downloads/m7yap%20dec%20july%2009.pdf.

Earth-Mates Dialogue Center, Kuwait Ministry of Awqaf and Islamic Affairs, and other participants. (n.d.). The Muslim seven year action plan on climate change 2010–2017—summary. https://iefworld.org/fl/WindsorARCMuslim_summary091020.pdf.

EcoMENA. (n.d.). Home page (contact and vision and mission). https://www.ecomena.org/.

Epley, J. L. (2010). Voices of the faithful: Religion and politics in contemporary Indonesia [Unpublished doctoral dissertation]. University of Michigan-Ann Arbor.

Gelling, P. (2009, November 16). Indonesia: The home of 'green Islam.' *GlobalPost*. https://www.pri.org/stories/2009-11-16/indonesia-home-green-islam.

Gottlieb, R. S. (2009). *The Oxford handbook of religion and ecology*, online ed., s.v. Introduction: Religion and ecology—What is the connection and why does it matter? Oxford University Press. http://www.oxfordhandbooks.com/view/10.1093/oxfordhb/9780195178722.001.0001/oxfordhb-9780195178722-e-1#oxfordhb-9780195178722-note-2.

GreenFaith. (n.d.). Campaigns and mobilizations. https://greenfaith.org/campaigns.

Green Muslims. (n.d.). Home page. https://www.greenmuslims.org/.

Inglehart, R., Haerpfer, C., Moreno, A., Welzel, C., Kizilova, K., Diez-Medrano, J., Lagos, M., Norris, P., Ponarin, E., & Puranen, B., et al. (Eds.). (2014). *World Values Survey: Round Six-Country-Pooled Datafile*. JD Systems Institute. http://www.worldvaluessurvey.org/WVSDocumentationWV6.jsp.

Interfaith Power & Light. (2017, June). Fact sheet—2017. https://www.interfaithpowerandlight.org/wp-content/uploads/2017/06/IPL-Fact-Sheet-June-2017.pdf.

International Islamic Climate Change Symposium. (2015). Islamic declaration on global climate change. http://www.ifees.org.uk/wpcontent/uploads/2016/10/climate_declarationmMWB.pdf.

International Muslim conference on climate change pressures OIC to act. (2010, April 12). *Islam Today*. https://web.archive.org/web/20100418154310/http://en.islamtoday.net/artshow-235-3579.htm.

Khaleafa. (n.d.). Home page (about and green Khutbah Campaign). http://www.khaleafa.com/.

Khattak, M. K., Khan, M. Y., & Hayat, M. (2019). Islam and environment: Time for green jihad. *Pakistan Journal of Islamic Research, 20*(2), 97–106.

Noer, D. (1973). *The modernist Muslim movement in Indonesia: 1900–1942*. Oxford University Press.

Pew Research Center. (2017a, January 31). World's Muslim population more widespread than you might think. https://www.pewresearch.org/fact-tank/2017/01/31/worlds-muslim-population-more-widespread-than-you-might-think/.

Pew Research Center. (2017b, November 17). Interactive data table: World Muslim population by country. https://www.pewforum.org/chart/interactive-data-table-world-muslim-population-by-country/.

Pope Francis. (2015). Encyclical letter, on care for our common home. *Libreria Editrice Vaticana*. http://w2.vatican.va/content/francesco/en/encyclicals/documents/papa-francesco_20150524_enciclica-laudato-si.html.

The Noble Qur'an. (n.d.). Sahih International translation. https://quran.com.

Rahman, A. S. (2017, February). Religion and animal welfare—An Islamic perspective. *Animals 7*(2), 11. https://doi.org/10.3390/ani7020011.

Reichley, J. A. (1985). *Religion in American public life*. The Brookings Institution. Quoted in M. Corbett & and J. M. Corbett. (1999). *Politics and Religion in the United States*. Garland Publishing, Inc.

Smith, G. (2002, Summer). Islam and the environment. (World and the spirit: Animals and the Qur'an). *Earth Island Journal, 17*(2), 26.

United Nations Climate Change. "External Statement—Islamic Declaration on Climate Change" 2015. https://unfccc.int/news/islamic-declaration-on-climate-change.

Wihbey, John. "'Green Muslims,' Eco-Islam and Evolving Climate Change Consciousness." *Yale Climate Connections*, April 11, 2012. https://www.yaleclimateconnections.org/2012/04/green-muslims-eco-islam-and-evolving-climate-change-consciousness/.

Wike, R., & Castillo, A. (2018, October 17). Many around the world are disengaged from politics. *Pew Research Center*. https://www.pewresearch.org/global/2018/10/17/international-political-engagement/.

Yildirim, K. A. (2016). Between anti-westernism and development: political Islam and environmentalism. *Middle Eastern Studies, 52*(2), 215–232.

Chapter 5

Darkness in the Rage of Light

Gendered EcoGothic Landscapes of the American West

Suzanne Roberts

On a recent backpacking trip, I realized how the literature I have read and its representations of women in literature have colored my view of the wilderness; for women, being outside alone can feel downright Gothic—and what I mean by Gothic isn't what many people think when they hear the word—about gargoyles and soaring buttresses (though the term was borrowed from Gothic architecture) or teenagers with black nail polish. Historically speaking, the Gothic picks up where the pastoral leaves off. Many critics agree the "traditional" Gothic spans the years from Horace Walpole's *The Castle of Otranto* to the publication of Mary Shelley's *Frankenstein* and Maturin's *Melmoth the Wanderer*. Nonetheless, Gothic tropes (e.g., distressed heroine, a villain who threatens the heroine's virginity, a crumbling castle or mansion, and frightening natural landscapes), as well as Gothic themes (e.g., isolation, insanity, and transgression fueled by desire) appear in literature spanning from the Graveyard Poets to the British Victorians, from nineteenth-century New England (Emily Dickinson and Edgar Allan Poe) to the twentieth-century American South (William Faulkner and Flannery O'Connor) to contemporary MTV video (Michael Jackson's *Thriller*) to present-day film.[1] Therefore, the pastoral and Gothic both function as modes of literature that span years and cross cultures. I prefer the term "mode" here to genre, the latter defining literary "types"—fiction, poetry, drama, nonfiction, and the various subsets and combinations thereof, while the former refers to the use of common tropes in literature.

The pastoral and the Gothic also share a penchant for broad and sometimes slippery definitions. E. J. Cleary (1995) argues that the term "Gothic," though used as a subtitle to Walpole's *The Castle of Otranto*, was not employed as a literary term until the twentieth century. The term Gothic, of course, comes from the Germanic tribes of the Goths who destroyed classical Roman civilization in invasions between 376 and 410 AD. Because the Goths left no art or literature of their own and were known only for invasion and destruction, the term Gothic came to be known as anything barbaric or uncivilized.

John Ruskin's characteristics of Gothic architecture also function to describe Gothic literature. He argues in *On Art and Life*, that the Gothic contains the following elements: "Savageness, Changefulness, Naturalism, Grotesqueness, Rigidity, and Redundance" (Ruskin). Literary critics have noted many of these same elements in Gothic literature. According to David Punter (2001), Gothic literature came to be associated with anything medieval or the opposite of classical: "Where classical was well ordered, the Gothic was chaotic; where the classical was simple and pure, Gothic was ornate and convoluted; where the classics offered a world of clear rules and limits, Gothic represented excess and exaggeration, the product of wild and uncivilized, a world constantly tended to overflow cultural boundaries." Fred Botting (1996) writes that while the Gothic signifies excess, it also romanticizes the past and questions contemporary values. He claims that the main features of the Gothic include "wild landscapes, the ruined castles and abbeys, the dark, dank labyrinths, the marvelous, supernatural events, (and) distant times and customs" (Botting, 2001). Like Botting, most critics describe setting—both natural and artificial—as integral to the Gothic.

Within the context of this argument, I'm talking about the Gothic heroine of the late eighteenth and nineteenth centuries. The female characters of those Gothic novels by Ann Radcliffe, Matthew Lewis, and later the Brontë sisters, who suffer strange and horrifying ends after engaging with a natural landscape, which reinforces the traditional pastoral idea that nature is a male space. Our stories are important; representation in art and literature informs our worldview the way we see the world and ourselves in it. The way women and nature have been portrayed in literature reinforce our beliefs that wild spaces are not, in the words of Christina Rosetti's famous poem "The Goblin Market," "good for maidens," as it is the place where the maiden is under threat, not so much from the natural world itself, but from the stranger—or Gothic villain—lurking in the shadows. Certainly, the solution would be the removal of those strangers, but instead, these books—and society itself—has blamed the women for being out there in the first place. Our culture has ingrained these ideas into our collective consciousness, so we do not stop to question this depiction. I think we should.

After a long climb to the top of California's Muir Pass, I celebrated by stretching my sleeping pad out on the ground and promptly collapsing on it. I leaned against an outcropping of rocks and looked out over Wanda Lake. Diamonds of afternoon light flashed on the surface of the water, recreating sky. John Muir says, "The Sierra should be called, not the Nevada or Snowy Range, but the Range of Light." I watched the sun sink below the ridgeline and the mountains change color—tan to yellow, Sierra alpenglow pink to the deepest purple. The surrounding valley unfolded below, glorious in its flood of light, but at the same time, the craggy granite towered above like a great Gothic cathedral or castle, so like the ones I had read about. The interplay between light and shade shifted the granite outcrops into shapes of flying buttresses, gargoyles, other creatures, reminding me of John Ruskin's notion that the Gothic cathedral takes its shape from the natural world. He says, "The original conception of Gothic architecture had been derived from vegetation—from the symmetry of avenues, and the interlacing of branches" (Ruskin). Although the castle is a symbol of artifice, and synecdochically the patriarchal Gothic villain, its form is derived from nature. And though the villain tends to be more closely related to culture (usually in the form of a duke or king or some other man in a position of power) rather than nature, it is in nature that the heroine is most afraid of him. Therefore, the Gothic castle borrows from the sublimity of nature in order to scare the Gothic heroine, reinforcing the idea that women ought to be afraid of the out-of-doors.

The terror of the mountains inspires awe, and the Gothic villain borrows from that awe-inspiring terror in order to frighten the heroine, making her fear both him and nature, but the terror cannot be too close, or it will not invoke the sublime. Burke says, "When danger or pain press too nearly, they are incapable of giving any delight, and are simply terrible." In truth, I have found much delight and peace in the mountains, but at the same time, sun-filled days fold into dusk, and night gives way to forest shadows and noises; moon shadows of the great pine and sounds of mice feet become, in my imagination, the bloodthirsty boogieman.

Maybe the wild things creep into my mind because I have been told that I, as a woman, should not be out in the forest alone after dark; after all, look what happens to the Gothic heroines after a foray into the woods. One must only think of those goblin men in Rossetti's famous poem or the count draining Lucy's blood in *Dracula* to be discouraged from wandering around the natural world alone. I know more women than men who find the wilderness foreboding. A study conducted by the National Recreation and Park Association found that "33 out of 35 women interviewed claimed that they would never hike alone in the woods because they would feel isolated and vulnerable to attack by a man" (Coble et al., 2003). I wonder if we, as American women, have inherited our strained relationship to the great outdoors from our pioneer foremothers, whose experiences were perhaps colored by their notions of the constructed wilderness? Constructs of wilderness have been created and reinforced by literature, especially the Gothic, which insists that the wilderness is necessarily frightening for women. The earlier generations who saw nature through the beautiful and sublime landscapes of the painters Claude Lorraine and Salvator Rosa created an inherited aesthetic of nature. The beautiful, pastoral landscapes of Claude and the sublime wilderness of Salvator, however, both represent male spaces. The British and American Gothic literature of the eighteenth and nineteenth centuries shows that neither space is open for women, and as a result, a woman's interaction with nature usually turns Gothic.

The literature of Ann Radcliffe draws heavily upon the notions of the beautiful and the sublime, and the Gothic heroine is always threatened when she ventures out into these landscapes. Radcliffe's heroines yearn for the outdoors, but the collective ideology that keeps women separate from nature always thwarts her efforts at a connection with nature. Finally, Radcliffe's heroines do settle into the pastoral landscapes, yet, they are always accompanied by a husband, showing that women really can't go out alone in nature. Radcliffe was ahead of her time in that her heroines did have adventures in the wilderness, yet, these forays into nature are always under extreme means; the heroine, then, remains blameless and is always restored to propriety. The poems inserted within Radcliffe's novels create women characters who are more able to break free from the expectations of gender. Because these poems are supposedly written by her characters, making them a sort of meta-speak, (women poets were not taken seriously, so when she presents poems, they are always written as if a character wrote them instead of being written by a female author) and they were largely ignored by her contemporary audience, Radcliffe is able to subversively deconstruct gender stereotypes, allowing the female to interact with the natural world. Even though she can't claim them for herself, Radcliffe is able to claim both nature and poetry for her female characters.

Because poetry was not, and is still not, profitable, women writers were able to write a poetry that challenges gender constructs in a way the novels, which were written for the market, could not. These authors allow their female speakers into pastoral and sublime landscapes. Although these landscapes are still foreboding and frightening, the heroines of their poems navigate this new wilderness in a way women in that time period could not. These poems, ahead of their time in many ways, could only exist as poems, a form not dictated by the pressures of the marketplace. With poetry, the woman is finally able to enter the male space of the pastoral landscape without becoming prey to bandits and villains, without becoming weak in both body and mind. Rather than the swooning Gothic heroine, the female shepherd is born; she acts as subject rather than object, placing her shepherd song alongside the rest.

Emily Brontë creates powerful fictional characters such as Catherine Earnshaw and Queen A.G.A. of the Gondal poems in order assert woman as the new female shepherd. Through these fictional characters, Brontë assumes the role of the female shepherd and her place within the pastoral landscape. Because of societal barriers, these pastoral moments lead to danger, including isolation, insanity, and death. Such fictional characters act as the forerunners to the feminist movement, able to do the things that the women of their time period only read about in books, such as traveling through the forest alone. Brontë herself confronts these tensions, resulting in an organic, wild writing style in both her novel and poems that blurs the line that separates humans from nonhuman nature.

The American poet Emily Dickinson is the literary heir to the Brontës.[2] Dickinson creates a poetry that defies tradition, and in doing so, she asserts herself as the female shepherd. She employs both Gothic and pastoral elements in her poetry in order to assert an environmental agenda, creating an ecogothic poetry. In order to live like the shepherd, Dickinson never marries, nor does she have children; rather, she becomes the "Wayward Nun," the "Myth of Amherst," or the pastoral shepherd, who tends her garden and sings her songs. Because Dickinson goes against the grain of her society, she is seen as strange. One neighbor calls Dickinson "the climax of the family oddity."[3] Faced with criticism and isolation, Emily Dickinson is the new pioneer of the literary frontier. Although Dickinson is able to assert herself as the female shepherd in her poems, real women of this era were still replicating the tropes of the Radcliffean heroine, especially women who had very real relationships to nature, such as those women who were following their husbands out West in covered wagons. In many respects, the American pioneer woman is the Gothic heroine of the westward expansion.

At about the time Muir was wandering through the range of light, families from the east were making their way across the continent in covered wagons. Between 1841 and 1866, approximately 350,000 people migrated westward (Holmes, 1995–2000). Many men, with gold on their minds and manifest destiny in their hearts, led their wives and children over 2,000 miles across the Overland Trail, finally climbing the steep, rocky ridge of the Sierra Nevada into California. For the pioneers, and especially the pioneer women, the Sierra Nevada was no range of light but a frightening, Gothic place. Although the Gothic conjures images of ominous castles and dark and stormy nights—the very antithesis of the wide-open space of a blue Sierra day—the Gothic is also defined by subversive mystery, new psychological and extrasensory pressures upon a harassed heroine, and the search for a hard-wrought happy ending (Summers, 1969). The pioneers, too, entered a mysterious place in search of that happy ending. And at the center of every Gothic novel is the heroine in distress. Both displaced and isolated, she is left to wander dark passageways. She must maintain her propriety, even while her virginity is constantly threatened. The pioneer woman shares this fear, real or imagined, as she travels westward across the wilderness.

While the settings differ, the Gothic heroine and the pioneer woman share many of the same struggles, so looking at the pioneer woman in terms of Gothic tropes reveals not only her situation but also our own modern-day connection to her. In the same way that the Gothic heroine who swooned and fainted throughout the pages of late eighteenth- and early nineteenth-century novels defined the Victorian woman in England, the pioneer woman continues to occupy our consciousness in terms of the woman under threat. Although most of their fears were not realized, the fear itself is a psychological terror that modern American women have inherited, thus straining our relationship to the wilderness. According to Coble et al. (2003), "Women may crave solitude but many fear being alone in the landscape. Over and over, they tell me they feel vulnerable; they feel danger—not from the land but from men." Like the Gothic heroine, both nineteenth-century westering women and contemporary American women experience the fear of threat in the wilderness.

Just as the heroine is at the center of the Gothic, the woman is at the center of the pioneer experience. Traditionally, it is the man who creates the pioneer experience, and his relationship to the wilderness—one of exploration, cultivation, and exploitation—defines the American imagination and identity (Nash, 1967). For this reason, the statues and memorials of pioneer families show the man, rather than the woman, front and center. A statue close to where I live in Lake Tahoe, California commemorates a party that became infamous because they were held up in an early October snowstorm in 1847 and forced to winter in the mountains (Donner Memorial State Park, 2014). In order to survive, they resorted to eating their dead, which is precisely why we remember them; we always remember the grisly stuff. The Donner party statue is like many others that mark the western landscape—a man standing out in front, looking westward, and behind him, a woman with a couple of children in tow—the archetypical view of the pioneer family.

The pioneers have defined our American relationship to the wilderness, and the western wilderness in particular; however, that relationship was rendered differently for the woman because, as in the statues, she follows. In many cases, she didn't want to travel westward. Margaret Hereford Wilson wrote to her mother in 1850: "Dr. Wilson has determined to go to California. I am going with him, as there is no other alternative" (Schlissel, 1992). According to Lillian Schlissel (1992), few women made the trip enthusiastically and for good reason. She claims:

> Such women subscribed to the journey with obedience and considerable courage, but not by acquiescence of their spirits. The women observed the rituals

> of patriarchal deference . . . however powerful to attempt to revise history, the period of the Overland Trail migration (1840–60) produces overwhelming evidence that women did not greet the idea of going west with enthusiasm, but rather that they worked out painful negotiation with historical imperatives and personal necessity. (Schlissel, 1992)

The Overland Trail was arduous and ended with a seventy-mile climb over the Sierra. The ascent involved hoisting wagons over snow with chains, pulleys, and improvised winches, which left women to walk, often pregnant, often holding babies and children.

Although the trip was difficult for the men, the responsibility of cooking and caring for the children fell on the shoulders of the women, making them the center of the wagon train party. Martha Ann Morrison, a fourteen-year-old girl who crosses the Overland Trail in 1844, writes, "The men had a great deal of anxiety and all the care of their families, but still the mothers had the families directly in their hands and were with them all the time, especially in sickness. . . . It strikes me as I think of it now that Mothers on the road had to undergo more trial and suffering than anybody else" (Schlissel, 1992). Despite the dangers of childbirth, many of the women survived, but their husbands and children often died, leaving them widowed and heartbroken.

Like the Gothic heroine, the pioneer woman confronted grisly events on a daily basis. According to Schlissel (1992): "No one who reads the diaries of women on the Overland Trail can escape feeling the intensity with which the women regarded loss of life." Many women became so preoccupied by death, that they obsessively kept track of the graves they saw each day on the trail. One woman, Ceceila McMillen Adams, reports the following: "June 25: Passed 7 graves . . . made 14 miles . . . June 26: Passed 8 graves . . . June 29: Passed 10 graves . . . made 22 miles" (Schlissel, 1992). Another pioneer woman, Sarah Davis writes in her diary: "We have past six graves to day we past twe[l]ve more and one grave they had not put the body in yet [*sic*]" (Holmes, 1995–2000). With so much sickness and death, these matter-of-fact reports are prevalent in many women's diaries. For these women, the new Eden is anything but a paradise. Like the Gothic heroine, her pastoral return means fear, bodily harm, and death.

Although mostly imagined, threat presses heavily on both the Gothic heroine and the pioneer woman. For the Gothic heroine, the physical world, an outside villain, and her own impending sexual coming-of-age represent the feared "Other." Ellen Moers (1978) says the Gothic heroine is "simultaneously persecuted victim and courageous heroine." Likewise, the pioneer woman must be courageous in the face of fear and threat. The pioneer woman's virginity, or at least her propriety, is threatened or seemingly threatened by the presence of other male emigrants and Native Americans, though the latter fear is more borne from xenophobia and racism; for her, this, in addition to the mountains, disease, wild animals, and nature in general all represent "The Other." Even though research proves that while the Native Americans were only sporadically hostile and usually helpful during the most important years of emigration, they were universally feared, especially by women.[4] The fear of the unknown (and misunderstood), whether it be the rumor of Indian hostility or the spooky thing behind the proverbial black curtain (as in Radcliffe's *Mysteries of Udolpho*), create the frightening "other" for women, especially in the outdoors. These representations of women as prey color the pioneer woman's experience, making her more fearful, causing more suffering. The Gothic heroine and the pioneer woman are simultaneously persecuted victim (if only in their own minds) and courageous heroine. Because they are real women, facing very real hardships, there is less report of swooning and fainting in the diaries of the pioneer woman, unless the woman in question is sick with mountain fever or some other ailment.

Along with fear, isolation plays an important role in the Gothic, and the heroine is left alone in nature. The heroine is usually sequestered in a lonely turret of the crumbling castle. Alone, she spends many hours singing to herself, writing poetry, playing the lute, and venturing outside but only along an outer corridor. Nature both disturbs and consoles her. Emily, the heroine of Radcliffe's *The Mysteries of Udolpho* finds nature both terrifying and sublime: "She saw only images of gloomy grandeur, or of dreadful sublimity, around her; other images, equally gloomy and equally terrible, gleamed on her imagination." The passing clouds shroud and reveal the mountains, and with each glimpse of them, Emily becomes terrified. But at the same time, she is consoled, "Emily lost, for a moment, her sorrows, in the immensity of nature" (Radcliffe). As in Shelley's "Mont Blanc," the viewer gives the mountains meaning. Because Emily views the mountains at a distance, she is able to find them delightful; though terrifying, they are also sublime.

At times, the westering woman was also isolated the midst of the mountains—too close to find them sublime, she sees them as simply terrible. Mary A. Jones traveled to California with her husband in 1846. Once they arrived, she remembered her husband asking, "Have you seen anything so beautiful?" To this, she replied, "There was nothing in sight but nature. Nothing . . . except a little mud and stick hut" (Holmes, 1995–2000). For her, nature and beauty were anything but synonymous; nature represented a difficult life. Because both pastoral and sublime landscapes have traditionally been European male spaces, the woman has been trained to look upon these landscapes with terror and therefore disdain.

When the pioneer woman does yearn for the pastoral, like the Gothic heroine, the situation often ends badly. Another emigrant woman, Mrs. Elizabeth Markham, initially describes the West as a forgiving, temperate place, yet as she travels westward, the land turns unforgiving and harsh: "o'er sinking sands and barren plains/our frantic teams would bound—/ While some were wounded, others slain" (Holmes, 1995–2000). The barren, sinking land acts as antagonist, causing injury and death. The Gothic and pioneer women both define and are defined by the land. Both enter the wilderness, seeking a pastoral return, and both suffer horrifying experiences. Unlike the pioneer man, the woman has come into the wilderness as his "helpmeet." She acts as object, without agency, left to the whims of her husband amid a wild landscape.

Although conditioned by societal expectations of women and nature—that she will suffer psychic and physical threat—the women prove an ability to survive. Sometimes, however, this threat becomes so great that it causes extreme psychological pressure, resulting in temporary or permanent insanity. Because women yearn for a connection to the land, which is not available to them, they succumb to insanity. In the famous death scene, Catherine from *Wuthering Heights* begs Nelly to open the window, so she can experience the out-of-doors. She is not permitted the freedom to run around the moors as she did when she was a girl, so she rents open the pillows, pulls out the feathers, and experiences nature in this way. She was taught long ago—after a bulldog bites her—that wandering around in nature is dangerous. Therefore, her desire for the unattainable wilderness renders her insane.

Some of the diaries from the westward expansion read like Gothic novels. A missionary wife, Narcissa Whitman, who had lost her only child in an accidental drowning on the Overland Trail, clung to her daughter's body for four days before giving her up for burial (Schlissel, 1992). Also, Jean Rio Barker, who traveled westward in 1851, similarly begged to "retain the little body" of her son (Holmes, 1995–2000). These events are reminiscent of the scene in Matthew Lewis's *The Monk*, when Agnes clings to her dead baby in the catacombs of St. Clare. Likewise, the pioneer woman suffers isolation and extreme hardship while maintaining a sense of propriety, which can lead to insanity.

Respectability is paramount for both the Gothic heroine and the pioneer woman. In Gothic literature, the heroine must maintain the utmost propriety; Emily from *The Mysteries of Udolpho* worries about having her head properly covered even though she is in the forest camping with bandits. Similarly, the pioneer women wore inconvenient clothing in order to uphold propriety. According to Schlissel (1992), "In their steadfast clinging to ribbons and bows, to starched white aprons and petticoats, the women suggest that the frontier, in a profound manner, threatened their sense of social role and identity." Most pioneer women wore long, flowing dresses and only the women who were rich enough to buy the latest fashions could afford to wear bloomers underneath. Even though the men often appropriated clothing from the land, animal skins and such, the women dressed quite impractically and many looked like they were heading out to an afternoon tea party in New England. Because of their prescribed "social roles," the dress code for these women further hindered their experience in the wilderness.

Pioneer women were, in fact, so concerned about respectability that they never wrote about matters considered private, such as bathing or childbirth. Many women report having babies in transit, without ever mentioning pregnancy beforehand. Amelia Stewart Knight's diary contains not one word of her pregnancy, even though she begins her trip well into her first trimester. When she gives birth, she says, "A few days later my eighth child was born. After this, we picked up and ferried across the Columbia River, utilizing skiff, canoes and flatboat to get across, taking three days to complete" (Schlissel, 1992). There is no mention of being pregnant, and the childbirth occupies only one sentence. The rest of the passage describes crossing a river but never mentions that Stewart has just given birth and must be carrying a newborn. Pregnancy is considered too private a subject even for a personal diary. Rarely do frontier women reveal how they feel about pregnancy and childbirth. Even though they were outside their constructed social space, taste in manners, social status, and respectability remain touchstones for the pioneer woman, further linking her to her literary predecessor, the Gothic heroine.

The Gothic heroine, of course, is a construct, a fictional character, and the pioneer women are actual people out in real wilderness landscapes. These diarists, however, inherit their notions of nature from their milieu, including literature. Descriptions of the landscape echo those that Radcliffe uses, the typical diction employed to describe nature by women and men alike in the eighteenth and nineteenth centuries. This shared vocabulary creates a lens through which to view the landscape. In a very real way, the metaphors become their lived reality. Although there is no mention of Radcliffe in her diary, Margaret A. Frink writes that the mountains around Lake Tahoe are "granite jaws" (Holmes, 1995–2000). She then describes how they pass into a "beautiful valley, sprinkled over with trees" (Holmes, 1995–2000). Certainly, the mountains surrounding Hope Valley must have seemed treacherous on their journey through them in 1850, but the description of "the jaws" of the mountains opening to a "sprinkling" of trees is reminiscent of the sublime landscapes of Salvator and the beautiful, pastoral landscapes of Claude, the painters Radcliffe relies on for her descriptions of the Alps. Sarah Raymond Herdon, who crossed the plains in 1865, used the language of the sublime and the beautiful in her descriptions of the Bear River Mountains in Idaho:

> The terrific storm, the broad prairies, the majestic forest, excite within our bosoms of awe and admiration, yet there are no places on earth that I have seen which have a tendency to inspire me with such tender feelings, such elevated,

> pure, holy thoughts as mountains. . . . Behold the mountains as they stand upon their broad bases, contemplate them as they rear their snowy tops in awful, majestic grandeur above the clouds, view then as you will, and they ever present the same untiring pleasure to the mind.

This description is uncannily similar to the opening passage in Radcliffe's (1998) *Mysteries of Udolpho*:

> It was one of Emily's earliest pleasures to ramble among the scenes of nature; nor was it in the soft and glowing landscape that she most delighted; she loved more the wild wood-walks, that skirted the mountain; and still more the mountain's stupendous recesses, where the silence and grandeur of solitude impressed a sacred awe upon her heart, and lifted her thoughts to the GOD OF HEAVEN AND EARTH.

Because Sarah Raymond Herdon was quite literate and later writing for publication, it is likely she knew some of the literature of her time, possibly including the novels of Ann Radcliffe. Like Radcliffe's natural landscape, the pastoral in Herdon's description foregrounds the sublime wilderness, inspiring feelings of awe and holiness in the speaker. Also, Herdon makes use of Radcliffean adjectives and nouns, such as "majestic" and "grandeur," which Radcliffe uses liberally in her novels to describe natural landscapes.[5] Herdon's view of the landscape in the nineteenth century was so ingrained into the collective imagination that even the "real" mountains become part of the imagined landscape of the sublime and the beautiful, proving that ideas of nature define the landscape, no matter if we are creating a fairytale of Gothic romance or describing a camping trip in Idaho. These literary constructs of pastoral and Gothic landscapes are superimposed on real landscapes, dictating the relationship between the human and her natural environment. The female Gothic shows that women desire an interaction with the natural landscape, because it is "awe-inspiring." However, this relationship traditionally ends in frightening consequences for the heroine. Although Emily Dickinson herself does not explore the wilderness, her poetry imagines a world where it is possible for a woman to be at home in the wild. Her poetry includes female speakers and characters who push the boundaries, safely inhabit wild places, and draw new maps of the world.

Therefore, in a very real way, the Gothic heroine sets the standard for the nineteenth-century lady. Supposed to obey rules of decorum, she only wanders into the wilderness in the most extreme circumstances. She was to protect her virginity at all costs, stay sweet, swoon at the right time, and marry the hero. The representations of the man in nature and the domestic woman in the house have had an impact on both men and women's relationships to the wilderness.

Men are also expected to fulfill a certain role in nature, and in contrast to women, they are never supposed to be afraid. Perhaps this fear is also the reason that "the highest risks to life seem to have been borne by the men" and pioneer women out-survived her male counterpart by two to one (Schlissel, 1992).

These Victorian ladies were expected to obey their husbands, fulfill their duty, and wander westward into the wilderness. Yet, they were expected to do so without breaking the rules of social conduct. They survived rain, wind, extreme temperatures, the death of loved ones, and subsequently, they quickly learned to hate the out-of-doors. According to Julie Roy Jeffery (1979), "By the 1840s, many middle-class women had internalized cultural ideals and norms of female behavior that were ill-suited to the realities of frontier life." These "cultural ideals" came, in part, from literature, and the Gothic novel had a large female readership. Perhaps these frontier women had even read these novels before leaving home; or, at least they were familiar with their themes and set of social constructs, especially pertaining to women and nature.

Both prototypes, the Gothic heroine and the pioneer woman, occupy the same place in our collective consciousness—they symbolize the woman under threat. The wilderness functions as a testing ground for both Gothic heroine and pioneer woman. And we hold our breath and wait to see what happens. It is not the actual wilderness, but the idea of it, passed down to us from our Gothic and pioneer foremothers, that frightens us. We cling to these notions because we never had a Danielle Boone; we can't claim a Jane Muir, so we turn to the heroine in the catacombs holding a candle, the woman walking alongside the covered wagon, carrying her newborn babe, and it is they who have defined for us our relationship to the wilderness. Even if we try not to believe it, we are told and retold both subliminally and overtly that when we venture into the wilderness, we will be threatened.

What frightens me most about the out-of-doors, it is the very thing most terrifying to our Gothic and pioneer foremothers—the idea of crazy man out there, the banditti or villain, the goblin or the vampire, threatening the maiden. But when I take a step back, or flop down on my sleeping pad in the late afternoon sun, gaze across the grand mountains, I can see them at once beautiful and terrible, pastoral and sublime. Although the mountains have been a place of darkness in our collective female past, we can break free from that notion, claim the natural landscape as a women's space, shepherd that terrain as our own, and see it the way Muir does, as the vast range of light that it is.

NOTES

1. Critics argue about the Gothic nature of present-day film as evidenced in the contradictory sentiments regarding Francis Ford Coppola's *Dracula*. Misha Kavka

in "The Gothic on Screen" *The Cambridge Companion to Gothic Fiction* (2002) calls the film more Gothic than the novel because of its use of the uncanny. In *Gothic* (1996), Fred Botting asserts that the Gothic "died" in the late 1990s and that Coppola's *Dracula* provides evidence because "The closure of the film is also emptied of Gothic effect: no climax, no solution, no sacred or rational expulsion of mystery, terror or duplicity" (180). Botting, however, leaves the door open for more Gothic to come. He says that "With Coppola's Dracula, the Gothic dies . . . Dying, of course, might just be the prelude to other spectral returns" (180).

2. Please see the reference section for a list of works for all three sisters: Anne, Charlotte, and Emily.

3. In a letter from young Mabel Loomis Todd (who later are Austin Dickinson's mistress) writes to her parents, "I must tell you about the *character* of Amherst. It is a lady whom people call the *Myth*. She is a sister of Mr. Dickinson, & seems to be the climax of the family oddity" (Qtd. in Gilbert, "The Wayward Nun beneath the Hill," Juhasz 22).

4. According to Elliot West, "Barely 350 whites, about one tenth of the one percent of the emigrants, had died at the hands of Indians. By comparison about 425 Indians were killed by whites, a number that took a much more significant toll on their small communities" (viii). Qtd. in Kenneth Holmes, Vol. 10.

5. In *The Mysteries of Udolpho* alone, Radcliffe uses the word "majestic" fifteen times and "grandeur" forty-nine times in describing the landscape.

REFERENCES

Botting, F. (1996). *Gothic: Critical idioms*. Routledge Press.

Botting, F. (Ed.). (2001). *Essays and studies 2001: The Gothic*. D.S. Brewer.

Botting, F. (2001). Gothic darkly: Heterotopia, history, and culture. In D. Punter (Ed.) *A companion to the Gothic* (pp. 3–15). Blackwell Publishing.

Brontë, A. (1988). *Agnes Grey*. Penguin Books.

Brontë, A. (1996). *The tenant of Wildfell Hall*. Penguin Books.

Brontë, A., et al. (1997). *Best poems of the Brontë sisters*. Dover Publications.

Brontë, C. (1974). *Shirley*. Penguin Books.

Brontë, C. (1985). *Jane Eyre*. Octopus Publishing.

Brontë. C. (2001). *Villette*. Modern Library.

Brontë, C. (2005). *The professor*. Barnes & Noble Press.

Brontë, E. (1952). *The complete poems of Emily Jane Brontë*. C. W. Hatfield (Ed.). Columbia University Press.

Brontë, E. (1985). *Wuthering heights*. Octopus Publishing.

Burke, E. (1986). *A philosophical enquiry into the origin of our ideas of the sublime and beautiful*. J. T. Boulton (Ed.). University of Notre Dame Press.

Clery, E. J. (1995). *The rise of supernatural fiction: 1762–1800*. Cambridge University Press.

Coble, Theresa G., et al. (2003). Hiking alone: Understand fear, negotiation strategies and leisure experience. *Journal of Leisure Research, 35*(1), 1–22.

Donner Memorial State Park Brochure. (2014). California department of parks and recreation. https://www.parks.ca.gov/pages/503/files/DonnerMemorialSP_Web2014.pdf.
Herdon, S. R. (2003). *Days on the road.* Morris Book Publishing.
Holmes, Kenneth J., (Ed.). (1995–2000). *Covered wagon women: Diaries and letters from the western trails.* University of Nebraska Press.
Jeffrey, J. R. (1979). *Frontier women: "Civilizing" the West? 1840–1880.* Hill and Wang.
Maturin, C. R. (2001). *Melmoth the wanderer.* Penguin Books.
Moers, E. (1978). *Literary women.* Women's Press.
Muir, J. (1988). *The mountains of California.* Dorset Press.
Nash, R. (1967). *Wilderness and the American mind.* Yale University Press.
Punter, D. (1996). *The literature of terror: A history of Gothic fictions from 1765 to the present day.* Addison Wesley Publishing Company.
Punter, D. (1998). *Gothic pathologies: The text, the body, and the law.* St. Martin's Press.
Punter, D. (2001). *Companion to the Gothic.* Blackwell Publishing.
Radcliffe, A. (1826). On the supernatural in poetry. *The New Monthly Magazine and Literary Journal, 16*(1), 145–152.
Radcliffe, A. (1998a). *The Italian.* Oxford University Press.
Radcliffe, A. (1998b). *The mysteries of Udolpho.* Oxford University Press.
Radcliffe, A. (1998c). *The romance of the forest.* Oxford University Press.
Ratchford, F. E. (1941). *The Brontë's web of childhood.* Columbia University Press.
Ratchford, F. E. (1955). *Gondal's queen: A novel in verse.* University of Texas Press.
Rossetti, C. (1994). *Goblin market and other poems.* Dover Publications.
Ruskin, J. (2004). *On art and life.* Penguin Books.
Schlissel, L. (1992). *Women's diaries of the westward journey.* Schocken Books.
Shelley, M. (1994). *Frankenstein.* Dover Publications.
Summers, M. (1969). *The Gothic quest: A history of the Gothic novel.* The Fortune Press.
Walpoles, H. (2004). *The castle of Otranto.* Dover Publications.

Chapter 6

Food Sustainability for the Underprivileged

A Comparison of Nonprofit Group Activities in Four U.S. Cities

Camila Pombo

INTRODUCTION

Issues related to the changing climate have increased over the past few decades, with an ongoing debate over what could be done to minimize the human impact on the natural environment. Indeed, how we produce, distribute, and consume our food carries significant weight in this debate. There has been a lot of research done on food sustainability and on what minimizing our diets' ecological footprint would look like. Fresh and locally grown food products form an essential part of a sustainable food system, but access to these practices can be quite limited for some. According to recent numbers, around 54.5 million people in the United States live in food deserts—where access to fresh and healthy food is limited (USDA, 2019).

This chapter will focus on the efforts of nonprofit organizations to connect underprivileged communities to sustainable food products and practices, mitigating the issues related to food insecurity and sustainability. More specifically, the research looks at four major U.S. cities—Houston, Oakland, Chicago, and New York, comparing their respective organizations and their programs that address this issue. The research focuses on five organizations—Urban Harvest (Houston), Acta Non Verba (Oakland), Growing Homes & I Grow Chicago (Chicago), and Just Food (New York). By identifying and categorizing their programs and initiatives, we can get a glance of what has been done to approach food accessibility and sustainability and what might be missing. With the hopes of finding information that will be beneficial to future efforts, this chapter attempts to answer the following research question: how do nonprofit environmental organizations from four major U.S. cities expand underprivileged communities' access to food sustainability?

LITERATURE REVIEW

There have been multiple studies that explore the issue of food sustainability and what an environmentally friendly food system would entail. Due to the complexity of the problem, there is an ongoing debate on what are the most important factors to take into consideration and what are the best options in terms of economic viability, social justice, carbon dioxide emissions, deforestation, among others. However, there are several overlaps between these studies that signal an overall ideal path as to what food sustainability could look like. Many argue that cutting down the consumption of animal products and focusing on a more plant-based diet would have significant benefits for both the environment and the overall health of the population (Garnett et al., 2014; Sáez-Almendros et al., 2013). Others look at the importance of localism and the positive repercussions that this type of system would entail, both in a socioeconomic and environmental aspect (Allen, 2010; Curtis, 2003). Finally, the accessibility of a sustainable diet for everyone and not just a small minority has not been overlooked. Many academics in the field have argued that for a food system to be considered sustainable, all groups in a community need to be included and these new systems must address issues such as poverty, hunger, and racial marginalization (Allen, 2010; Allen et al., 1991; Bateman et al. 2014; Duell, 2013).

The idea of a socially just sustainable food system raises questions about the role food deserts play in this issue and how viable and accessible sustainable diets and food systems are to minority communities around the United States. There have been multiple studies that examine the presence of food deserts in the United States and find a correlation between access to food and other social and economic factors. Some researchers argue that economic class is a massive contributor to this issue, where low-income neighborhoods have, on average, 75 percent fewer supermarkets compared to middle-income neighborhoods (Powell, 2007). Other studies suggest that race is also a significant factor, where access to supermarkets and therefore fresh and more sustainable produce in black neighborhoods is 52 percent that of predominantly white neighborhoods (Morland & Filomena, 2007; Powell, 2007). Globally, other studies explore the cost of fresh and healthy food products, demonstrating how low-income families would have to spend almost half of their income as compared to middle-income families, who spend less than ten percent of their income for the same products (Barosh et al., 2014).

Accordingly, there have been other papers that analyze what has been done to tackle this issue and which propose what could be done in the future. Many academics argue that urban agriculture and small-scale farming could be the key to connecting minorities to sustainable eating (Ackerman et al., 2014; Alteri, 2009). Additionally, others explore the importance of community outreach and children-focused programs in promoting these sustainable practices and including minority groups (Carlsson & Williams, 2008; Feenstra, 2002). This research project aims to add to the conversation, centering on the role nonprofit organizations play in connecting minority communities to sustainable food practices in the United States.

METHODS

This study compares four nonprofit environmental organizations that deal with food sustainability in four major United States cities—Houston, New York, Chicago, and Oakland. The cities were chosen based on the presence of these nonprofit organizations, along with data regarding the area's racial distribution, poverty rates, and food desert proportions. The organizations being explored include Urban Harvest (Houston), Just Food (New York), I Grow Chicago and Growing Homes (Chicago), and Acta Non Verba (Oakland). By analyzing the materials available online, their efforts were compared and categorized by types of programs.

The demographic statistics serve as a basis to compare the cities' minority population density. According to recent data, all four cities have quite diverse populations, with minority groups making up around 71 percent of the population on average (U.S. Census Bureau, 2018). These four cities all have a much larger minority population compared to the national average, which shows how the access minority and underprivileged communities have to sustainable food practices is relevant in each area, and therefore, increases the presence of nonprofit organizations in this sector.

As for the economic factors, the average poverty rate between these four cities is approximately 19 percent, which is drastically higher than the national average standing at 13 percent (U.S. Census Bureau, 2018). This contrasting data demonstrates the presence of underprivileged communities and the economic disadvantages of these. The information gathered in this section showcases the evidence for why these cities have multiple nonprofit organizations that deal with food sustainability and accessibility and, therefore, justifies the selection of these for the comparative analysis.

As most urban areas in the country, these four cities have a high density of food deserts around underprivileged areas, where access to healthy and fresh food is unjustly limited (USDA, 2019). Figure 6.1 shows the maps of

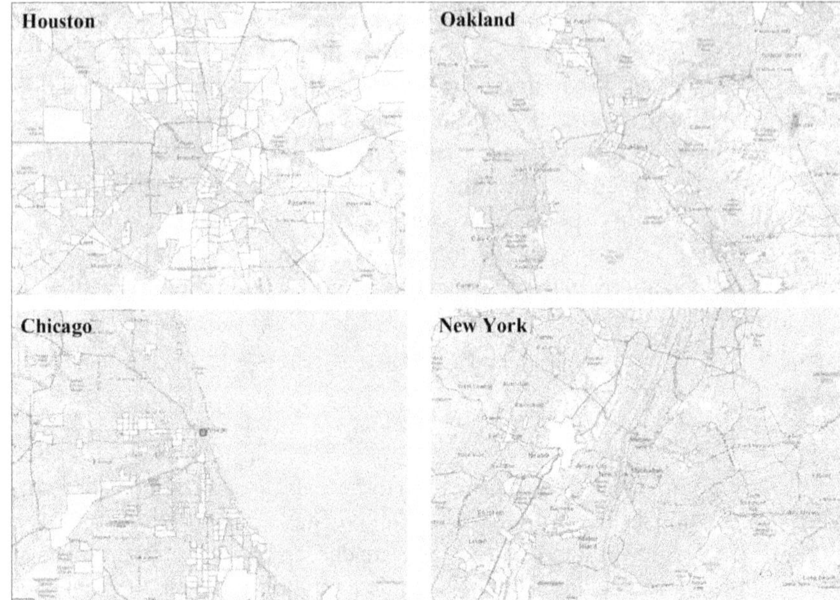

Figure 6.1 Food Desert Density on Each City. *Map Source*: U.S. Department of Agriculture, 2019

the cities and highlighted are the low-income areas where more than a 100 household units do not have a vehicle and are more than half a mile away from the nearest supermarket, exposing the high density of food insecurity and scarcity in these four cities.

As previously mentioned, multiple studies suggest that a sustainable food system is hugely based on access to fresh food, indicating that there is a gap between underprivileged communities and food sustainability in these areas and, therefore, denoting the need for nonprofit organizations that address this issue.

Each city has multiple nonprofit organizations that focus on food insecurity, food accessibility, and hunger. However, for this research, the organizations chosen are ones which have an environmental concentration as well, meaning that they take into consideration issues related to the conservation of the natural environment and the sustainable allocation of resources. The organizations chosen were Urban Harvest in Houston, Just Food in New York, I Grow Chicago and Growing Homes in Chicago, and Acta Non Verba in Oakland. For Chicago, two organizations were chosen since they collaborate and complement one another; additionally, together, they provide a more accurate representation of the city's efforts and therefore allow for a more precise and fair comparison with the other cities. For the purpose of clarity, these two organizations are counted as one in the comparative analysis.

This study uses an exploratory approach, where the information was gathered from the material available online for each organization. The types of documents used for this research include mission statements, fact sheets, annual reports, images, strategic plans, past interviews, and statistics available on their respective websites. For the comparative analysis, the organizations' initiatives were organized by the type of program, which include youth and community involvement, urban gardening and farming, and local economic empowerment. The information was then shown in a table format that allowed for a clearer comparison which helped to answer the research question presented: How do nonprofit environmental organizations connect with underprivileged communities on food sustainability in four major U.S. cities?

ANALYSIS

The organizations explored in this research have different programs, structures, and strategies; however, they all share the same focus: to address issues of food sustainability and accessibility in relation to minority communities in the area. The most relevant organization found in Houston is called Urban Harvest—a nonprofit organization that works on expanding access to fresh and sustainable food. According to their mission statement, the organization aims to "cultivate thriving communities, through gardening and access to healthy, local food." (Urban Harvest, 2019). They are a grassroots organization that started more than thirty years ago in the middle of the Fourth Ward—one of the most impoverished areas in Houston. According to their 2018 annual report, "[their] community garden program accounts for over 7 acres of diversified productive greenspace in urban areas [...] and spans over 100 miles across the greater Houston area" (Urban Harvest, 2018).

For Oakland, this study focuses on an organization called Acta Non Verba: Youth Urban Farm Project (ANV), whose primary goal is to increase living standards of minority communities in Oakland, CA, through the use of urban farming with a specific emphasis on children and women. According to their mission statement, the organization aims to "elevate life in Oakland and beyond by challenging the oppressive dynamics and environments through urban farming" (ANV, n.d). Additionally, they are the only organization in this study that also has a woman empowerment approach, where the organization itself was "founded and [is led] mainly by women of color from surrounding neighborhoods and the larger community" (AND, n.d).

In the Chicago area, there were two organizations chosen, as they collaborate and complement each other. Growing Home is a nonprofit organization that works to increase living standards for underprivileged areas in the city, through programs related to job training, urban food growing, and community-supported agriculture. According to their 2018 annual report, out of the program's participants, "86% of them went on to secure full-time employment." Additionally, the organization was able to produce 26, 217 pounds of certified organic product that was harvested and sold in the Chicago area, and of which 17 percent was donated and sold at a reduced price in food deserts (Growing Homes, 2018). For the other organization in Chicago, the research looks at I Grow Chicago, which is a smaller organization that focuses on the community in Englewood—a prime food desert in the city. According to their mission statement, their goal is "to grow Englewood from surviving to thriving through community connection, skill-building, and opportunity" (I Grow Chicago, 2020). They have a more community-centered and children approach, which results in smaller scale impacts.

Finally, the organization chosen for New York is called Just Food—which aims to empower marginalized communities and connect them to locally grown food and sustainable practices. They have a specific focus on leadership and youth involvement, where the organization aims to "support community leaders in their efforts to advocate for and increase access to healthy, locally grown food, especially in underserved NYC neighborhoods" (Just Food, 2015). Through their food market connections and workshops, they have managed to raise the number of New Yorkers that have access to fresh and sustainably grown food from 25,000 residents in 2003, to around 250,000 residents in 2015 (Just Food, 2015).

As mentioned earlier, there are differences among the organizations' programs, structures, approaches, and management styles. Some of the organizations center their efforts on children or women, others on employment and leadership. However, based on their mission statements and goals, they all seem to aim at expanding the access minority groups and underprivileged communities have to healthy and sustainable foods. In this research, we look specifically at their programs, categorizing them under three types—youth and community involvement, urban gardening and farming, and local economic empowerment. The following table includes the types of programs for all nonprofit groups discussed in the next sections.

Youth and Community Involvement

The inclusion of youth groups and community members in educational efforts can be essential in any type of cause. This category refers to programs that have a specific focus on either education or family outreach. All the organizations analyzed in this research have a program that falls under this group. Table 6.1 shows all the different programs for each organization, subdivided

Table 6.1 Nonprofit Programs Categorized

	Youth and community involvement	*Urban gardening and farming*	*Local economic empowerment*
Urban Harvest (HOU)	**Youth education** - gardening classes - school affiliations	**Community gardens** - community and affiliated gardens - Adopt-A-Garden	**Economic growth-** farmers' markets (meet your vendors)
	Community outreach - gardening classes for continuing education	**Gardening resources** - online resources - free classes - hub distributions	**Financial tools** - SNAP - Double-Up Houston
Acta Non Verba (OAK)	**Youth education** - Camp ANV - Tassafaronga Village	**Community gardens** - A's, ANV's and WOW urban farms	**Micro-economy** - community-supported agriculture
	Community outreach - Leaders in Training	**Gardening resources** N/A	**Financial tools** - East Oakland Grocery Cooperative
Growing Homes and I Grow Chicago (CHI)	**Youth education** - Peace Garden	**Community gardens** - Peace Garden - urban farms	**Micro-economy** - farmers' markets
	Community outreach - Family Resource Home	**Gardening resources** N/A	**Financial tools** - Job Training Model
Just Food (NY)	**Youth education** N/A	**Community gardens** N/A	**Micro-economy** - community-supported agriculture - farmers' markets (value chain map)
	Community Outreach - Just Food Conference - gardening classes - Youth Empowerment Pipeline	**Gardening resources** - free urban agriculture lessons	**Financial tools** - Just Share (SNAP) - pantries

Source: This table has been created by the author.

into two different sections: youth education and community outreach. It is important to mention that some of these programs may overlap, as they are not always independent of one another. However, since some organizations have very distinct programs in both categories, it is worth exploring them separately.

While most organizations have some sort of program affiliated with youth education, they are all relatively different. Urban Harvest provides gardening classes to kids through their after-school programs, where they develop outdoor classrooms for food growing. According to their latest annual report, around 11,000 students were reached through their efforts (Urban Harvest, 2018). Acta Non Verba has a program called Camp ANV, which is an eight-week camp that provides education related to business and sustainable agriculture (ANV, n.d). In Chicago, the organization I Grow Chicago has an initiative called the Peace Garden (which is located in the middle of Englewood) that provides urban gardening education to the kids that participate in their after-school programs (I Grow Chicago, 2020).

Under youth education, there are multiple programs described as after-school programs. Urban Harvest has created school partnerships with twenty-three elementary schools in the Houston area, providing after-school urban gardening classes in underprivileged areas (Urban Harvest, 2019). In Oakland, Acta Non Verba has an affiliation with the Tassafaronga Village After School Program, where the organization provides low-income households the opportunity to participate in the program free of charge (ANV, n.d). Finally, the organization I Grow Chicago brings activities for kids in their Peace Campus that include yoga, creative arts, and, as previously mentioned, urban gardening classes (I Grow Chicago, 2020).

While the previous section is more centered toward younger kids, other programs are specifically centered on leadership training, falling under the community outreach efforts. Acta Non Verba has a program called Leaders in Training (LIT) that provides teenagers the opportunity to become better leaders in their communities, with a focus on sustainable and healthy living. This program also allows for young teenagers to possibly be offered a job after completion of the curriculum, presenting low-income families with solutions to avoid the poverty trap (ANV, n.d). Moreover, Just Food in New York brings the opportunity for young adults to participate in their program called Youth Empowerment Pipeline—which aims to provide them with experience in the social justice movement. In conjunction with their advocacy training workshops, this program looks specifically at issues related to food justice and food sustainability (Just Food, n.d).

Since education is not limited to children, the community outreach programs also explore continuing education efforts. Every year, Urban Harvest offers approximately sixty gardening classes in subjects related to urban gardening and sustainability, including permaculture, fruit tree caring, and sustainable living. These classes may sometimes count for continuing professional education credits by meeting the Texas Education Agency requirements (Urban Harvest, 2019). Similarly, Just Food in New York also offers multiple workshops and classes for the community gardeners and urban farmers. According to their 2015 annual report, the organization led 41 workshops for nearly 500 different neighborhoods. Additionally, the organization also coordinates its annual Just Food Conference, where a variety of topics related to food justice and sustainable urban food systems are explored (Just Food, 2015). Apart from their youth-centered programs, I Grow Chicago has another program called the Family Resource House, where they provide support to parents, especially women, and address issues related to financial stability and domestic violence (I Grow Chicago, 2020).

Urban Gardening and Farming

Food sustainability is highly related to both the production and consumption of locally grown food, where urban gardening and farming plays an essential role. This category refers to programs that mainly focus on the growing and production of fresh food, including any production scale. Since all the organizations center their efforts on food sustainability, it comes with no surprise that all have programs that fall under this category. In Table 6.1, the different programs for each city's organization are listed and divided into two sections—community gardens and gardening resources availability. Similar to the previous section, there is a lot of overlap between these two categories, but it is vital to distinct them, as some have different goals and methods. Also, there is an overlap in some programs with the other two types of program categories, as they are, in a way, all interconnected.

As shown in table 6.1, three out of the four organizations have some sort of community garden program, expanding residents' access to sustainable food and urban gardening. Urban Harvest has in total 140 affiliate gardens, out of which 65 are located in areas designated as food deserts, and 25 of them donate all their produce to shelters, pantries and meal centers in the area. Additionally, Urban Harvest also includes a program called Adopt-A-Garden, which connects gardens in food deserts with community partners to revitalize them and ultimately turn the space into a food garden for the neighborhood (Urban Harvest, 2019). In relation to urban food growing, Acta Non Verba has in total three different locations where they promote urban gardening for food production. These three urban farms (A's, ANV's, and WOW farms) are also related to the after-school programs that they provide, allowing for kids to learn about sustainable food growing and consumption practices (ANV, 2018). For Chicago, the organization I Grow Chicago has its

Peace Garden that consists of a small neighborhood community garden where food growing is encouraged as part of the activities for children. The other Chicago organization, Growing Homes, has two central urban farms called Les Brown Memorial Farm and Wood Street Urban Farm. Both, which will be explored later on (Growing Homes, n.d).

The second section of this category refers to programs that facilitate urban gardening, giving communities the resources needed to grow fresh and healthy food successfully. Through their urban gardening programs, Urban Harvest allows for members that either work with an affiliate garden or with their Adopt-A-Garden program to access their classes for free. Additionally, they give out gardening supplies and materials to these community gardens through their hub distribution locations and through their vast online platform, which includes documents regarding gardening seasonal sheets, types of fruits and vegetables, and gardening instructions (Urban Harvest, 2019). Similarly, Just Food in New York also provides urban agricultural classes and workshops to help community gardens thrive and successfully offer fresh produce to the surrounding areas (Just Food, n.d.).

Local Economic Empowerment

Economic stimulation is a significant component of groups that address any socioeconomic issue. This category includes programs that are primarily connected to the advancement of economic development in each community, by either stimulating micro-economic systems (i.e., farmers' markets) or by providing residents with financial tools. Table 6.1 includes all the programs that pertain to this category, divided by the two different sections previously mentioned. The financial tools section includes programs that take a direct or indirect approach. Direct financial tools refer to programs that provide immediate help, such as money transactions or food resources. Indirect financial tools relate to programs that give skills and/or tools that the residents will be able to utilize to later on get some financial benefit on their own.

The section Economic growth refers to programs that create or boost economic activity for underprivileged small communities. As shown in Table 6.1, every organization has some sort of program that falls under this category, and all of them are related to sustainable food production or consumption. In Houston, Urban Harvest provides a platform for local food producers and connects them with the communities around them. They organize three markets every week around the city area. According to their reports, they support over seventy local Houston farmers' market vendors (Urban Harvest, 2019). Similarly, Just Food in New York connects residents with twenty-seven community-led farmers' markets through their network hubs and online interactive map. Also, they facilitate a community-supported agriculture system (CSA) where residents give early bulk payments allowing for farmers to successfully produce every season (Just Food, n.d.). Likewise, Acta Non Verba in Oakland also promotes a CSA system through food production in their WOW urban farm, whose products are only sold to Oakland residents and businesses (ANV, n.d.). Finally, in Chicago, Growing Homes sells all of their fresh food produced by their two high-production urban farms to people within a 20-mile radius. According to their 2018 annual report, around 4,300 pounds of fresh organic produce was sold in Englewood alone, which, as mentioned before, is one of the biggest food deserts in the Chicago area (Growing Homes, 2018).

Indirect financial tools allow for the participants to earn valuable skills that will later help them find jobs and become financially stable. The primary focus of Growing Homes is to improve the lifestyle of marginalized communities in the city, and they do this mainly through their Job Training Model program (indirect financial tool program)—where low income and, sometimes, previously incarcerated individuals have the opportunity to develop their strengths as employees. Their urban farms serve as a training model, with the hopes of helping these individuals find secure employment after the completion of the program (Growing Homes, n.d.). Moreover, Acta Non Verba has a program still in the process of development called East Oakland Grocery Cooperative, which aims to provide fresh and healthy food and job opportunities for the racial minority groups in Oakland (ANV, n.d.). While it seems quite promising, the program is still not active.

On the other hand, direct financial tool programs aim to provide immediate financial relief to the families and residents around the areas. Both Urban Harvest in Houston and Just Food in New York collaborate with governmental initiatives such as Supplemental Nutrition Assistance Program (SNAP) to allow for low-income households to purchase fresh food for a lower price. In Houston explicitly, Urban Harvest farmers' markets accept Double-Up Food Bucks, which gives individuals a dollar-for-dollar match on benefits for fresh food produce (Urban Harvest, 2019). Similarly, Just Food has a program called Just Share, which works to make CSA membership more affordable to low-income households by matching payments with SNAP benefits (Just Food, 2015).

DISCUSSION

The nonprofit organizations of four U.S. cities were analyzed and categorized by their types of programs. Even though they all share the same goal of connecting underprivileged communities with food sustainability efforts, they vary in structure and strategies. Based on their programs and mission statements, they all have a social focus, where the development and progression of minority and underprivileged communities in their respective areas is their number one priority. The findings suggest that, while different, these organizations overlap on three types of programs implemented: youth and community involvement, urban gardening and farming, and local economic empowerment. The organizations in Houston and Chicago had the most overlap between these categories and are therefore considered to be the ones with the most holistic approach.

For the youth and community involvement sector, we look at two different categories: youth education and community outreach. While most organizations have programs that strive to increase youth education, Just Food in New York is the only one that does not have a particular program specifically designed for younger kids. Also, the three organizations that have an after-school program (Urban Harvest, Acta Non Verba, and I Grow Chicago) all allow for low-income households to participate by having free of charge options. Moreover, programs related to leadership training do not seem to have the same level of importance as only two of the four organizations have programs with this focus. Overall, the information gathered in this category suggests that educational involvement (for both young kids and older community members) is an essential part to successfully include minority communities in food sustainability efforts.

As for the urban gardening and farming category, all organizations studied have a deep focus on community gardens and food production. Locality is an essential part of food sustainability, so it is not surprising that all of these organizations had a program related to this concept. Every organization, except for Just Food in New York, has an initiative that provides a space for food production, through the creation of community gardens or urban farms. The second subcategory refers to the organizations expanding the availability of gardening resources. While most of them do promote urban gardening, only Urban Harvest in Houston and Just Food in New York actually provide both learning and gardening resources to facilitate these efforts. In comparing all the programs, Urban Harvest in Houston appears to be the organization with the most complete approach, as they are the only organization that provides both a space for urban gardening and the necessary gardening resources.

While environmental sustainability is an essential part of their platforms, the organizations' primary emphasis is on the communities' well-being rather than on the overall environmental benefit, which comes as a close second. For the last category, the findings suggest that all organizations have some sort of program that provides a closed-loop economic system revolving around sustainable food (i.e., fresh locally grown produce). Judging from the structure of the programs and the information available for the public, there seems to be a more reliable connection between the vendors at the farmers' markets and the organization in Houston. As for accessibility, both Urban Harvest in Houston and Just Food in New York have a direct financial assistance method, were they both work with governmental programs, such as food stamps, that help decrease the food cost for low-income families. On the other hand, Growing Homes in Chicago has a more indirect financial assistance approach, where they provide job training, which aims to later on provide a greater financial relief. As for Acta Non Verba in Oakland, their program East Oakland Grocery Cooperative would be under this category, but the program is still under development.

The organizations in Houston and Chicago seem to achieve the most rounded approach, based on their two programs: Adopt-A-Garden from Urban Harvest in Houston and the Job Training Model from Growing Homes in Chicago. Even though both were categorized under just one type of program, they seem to include aspects from all three categories. Adopt-A-Garden in Houston concretely addresses communities located in food deserts, and the organization's hub distributions, online resources, and educational programs all allow for these communities to successfully maintain a high-yielding community garden. On the other hand, the Job Training Model from Growing Homes in Chicago also covers all three sectors, by allowing for underprivileged individuals to access educational tools and labor training through urban gardening processes, with the idea of creating economic development in the most affected communities.

Through a comparison of groups in four major U.S. cities, this research finds that nonprofit organizations connect with minority communities on food sustainability through the implementation of three types of programs: youth and community involvement, urban gardening and farming, and local economic empowerment. The most holistic programs were found in Houston (Adopt-A-Garden) and Chicago (Job Training Model). Future research should be done in the organization's impact on the communities through mapping and census reports. Also, this research only focused on major urban areas, so future research should also include an analysis of more rural areas and what has been done there. One of the limitations of this research is that the type of information available was different for each organization, creating some inconsistencies within the findings. Finally, all of these nonprofit organizations only work on a local scale, so there is still the question of what can be done on a national or even global scale to make our food system more sustainable and accessible. Food sustainability should be for everyone, not just the ones who can afford it.

REFERENCES

Ackerman, K., Conard, M., Culligan, P., Plunz, R., Sutto, M. P., & Whittinghill, L. (2014). Sustainable food systems for future cities: The potential of urban agriculture. *The economic and social review*, *45*(2, Summer), 189–206.

Acta Non Verba. (2018). Acta Non Verba Partners with the Oakland Athletics for "The Farm!". https://anvfarm.org/athleticsfarm/

Acta Non Verba. (n.d.). Acta Non Verba: Youth Urban Farm Project. https://anvfarm.org/

Acta Non Verba. (n.d.). Camp ANV's 2020 Camps. https://anvfarm.org/programs/camp-anv/

Acta Non Verba. (n.d.). East Oakland Grocery Cooperative. https://anvfarm.org/east-oakland-grocery-cooperative/

Acta Non Verba. (n.d.). Leaders in Training (LITs). https://anvfarm.org/programs/camp-anv/leaders-in-training/

Allen, P., Van Dusen, D., Lundy, J., & Gliessman, S. (1991). Integrating social, environmental, and economic issues in sustainable agriculture. *American Journal of Alternative Agriculture*, *6*(1), 34–39.

Allen, P. (2010). Realizing justice in local food systems. *Cambridge Journal of Regions, Economy and Society*, *3*(2), 295–308.

Alkon, A. H. (2012). *Black, white, and green: Farmers markets, race, and the green economy*. Athens: University of Georgia Press.

Altieri, M. A. (2009). Agroecology, small farms, and food sovereignty. *Monthly Review*, *61*(3), 102–113.

Barosh, L., Friel, S., Engelhardt, K., & Chan, L. (2014). The cost of a healthy and sustainable diet–who can afford it? *Australian and New Zealand Journal of Public Health*, *38*(1), 7–12.

Carlsson, L., & Williams, P. L. (2008). New approaches to the health promoting school: Participation in sustainable food systems. *Journal of Hunger & Environmental Nutrition*, *3*(4), 400–417.

Curtis, F. (2003). Eco-localism and sustainability. *Ecological Economics*, *46*(1), 83–102.

Duell, R. (2013). Is local food sustainable? Localism, social justice, equity and sustainable food futures. *New Zealand Sociology*, *28*(4), 123.

Feenstra, G. (2002). Creating space for sustainable food systems: Lessons from the field. *Agriculture and Human Values*, *19*(2), 99–106.

Garnett, T., Appleby, M. C., Balmford, A., Bateman, I. J., Benton, T. G., Bloomer, P., & Herrero, M. T. (2014). What is a sustainable healthy diet? A discussion paper.

Growing Homes. (2018). Annual Report 2018. http://growinghomeinc.org/docs/GrowingHome-AnnualReport_FY2018.pdf

Growing Homes. (n.d.). Our Farms. http://growinghomeinc.org/our-farms/

Growing Homes. (n.d.). Our Job Training Model. http://growinghomeinc.org/our-model/

I Grow Chicago. (2020). Peace Campus. https://www.igrowchicago.org/peace-campus/

I Grow Chicago. (2020). Our Work. https://www.igrowchicago.org/our-work/

Just Food. (2015). Annual Report 2015. http://www.truevaluemetrics.org/DBpdfs/Initiatives/JustFood/Just-Food-Annual-Report-2015.pdf

Just Food. (n.d.). Value Chain Map. https://www.justfood.org/value-chain-map

Just Food. (n.d.). Youth Empowerment Pipeline. https://www.justfood.org/youth empowermentpipeline

Maćkiewicz, B., Asuero, R. P., & Almonacid, A. G. (2019). Urban agriculture as the path to sustainable city development. Insights into allotment gardens in Andalusia. *Quaestiones Geographicae, 38*(2), 121–136. https://doi-org.ezproxy.lib.uh.edu/10.2478/quageo-2019-0020

Morland, K., & Filomena, S. (2007). Disparities in the availability of fruits and vegetables between racially segregated urban neighbourhoods. *Public Health Nutrition, 10*(12) , 1481–1489.

Mui, Y., Ballard, E., Lopatin, E., Thornton, R. L. J., Pollack Porter, K. M., & Gittelsohn, J. (2019). A community-based system dynamics approach suggests solutions for improving healthy food access in a low-income urban environment. *PLoS ONE, 14*(5), 1–13. https://doi-org.ezproxy.lib.uh.edu/10.1371/journal.pone.0216985

Powell, L. M., Slater, S., Mirtcheva, D., Bao, Y., & Chaloupka, F. J. (2007). Food store availability and neighborhood characteristics in the United States. *Preventive Medicine*, *44*(3), 189–195.

Ritchie, H., & Roser, M. (2020). Environmental impacts of food production. *OurWorldInData.org*. https://ourworldindata.org/environmental-impacts-of-food

Sáez-Almendros, S., Obrador, B., Bach-Faig, A., & Serra-Majem, L. (2013). Environmental footprints of Mediterranean versus Western dietary patterns: beyond the health benefits of the Mediterranean diet. *Environmental Health*, *12*(1), 118.

Sublette, C., & Martin, J. (2013). Let them eat cake, caviar, organic, and whole foods: American elitism, white trash dinner parties, and diet. *Studies in Popular Culture, 36*(1), 21–44. www.jstor.org/stable/23610150

Taylor, D. (1997). American environmentalism: The role of race, class and gender in shaping activism 1820–1995. *Race, Gender & Class, 5*(1), 16–62. www.jstor.org/stable/41674848

United States Census Bureau. (2018). ACS Demographic and Housing Estimates. https://data.census.gov/cedsci/table?q=United%20States&tid=ACSDP1Y2018.DP05&g=0100000US&hidePreview=true&table=DP05

United States Census Bureau. (2018). Quick Facts: United States. https://www.census.gov/quickfacts/fact/table/US/PST045219

United States Department of Agriculture. (2009). Access to affordable and nutritious food: Measuring and understanding food deserts and their consequences. Economic Research Service. https://www.ers.usda.gov/webdocs/publications/42711/12716_ap036_1_.pdf

United States Department of Agriculture. (2019). Food Access Research Atlas. Economic Research Service. https://www.ers.usda.gov/data-products/food-access-research-atlas/go-to-the-atlas/

United States Department of Agriculture. (2019). Food Access Research Atlas: Documentation. Economic Research Service. https://www.ers.usda.gov/data-products/food-access-research-atlas/documentation/

Urban Harvest. (2018). Annual Report 2018. https://www.urbanharvest.org/about/annual-report/

Urban Harvest. (2019). Community Gardens: Resources & Gardening Advice. https://www.urbanharvest.org/gardening-advice/

Urban Harvest. (2019). Education & Classes: Edible Academy. https://www.urbanharvest.org/education/edible-academy/

Urban Harvest. (2019). Education & Classes: School Partnerships. https://www.urbanharvest.org/education/youth/school-partnerships/

Urban Harvest. (2019). Education & Classes: School Gardening Resources. https://www.urbanharvest.org/education/youth/resources/

Urban Harvest. (2019). Houston Farmers Markets. https://www.urbanharvest.org/farmers-market/

Urban Harvest. (2019). SNAP Benefits. https://www.urbanharvest.org/farmers-market/SNAP/

Urban Harvest. (2019). Urban Harvest: Our Mission. https://www.urbanharvest.org/about/

Walker, R. E., Keane, C. R., & Burke, J. G. (2010). Disparities and access to healthy food in the United States: A review of food deserts literature. *Health & Place, 16*(5), 876–884.

Chapter 7

Egalitarians Speak

Lone Voices of Dissent in the Congressional Hearings on Radical Animal Liberation and Environmental Activism

John A. Duerk

INTRODUCTION

Radical activism in the United States has roots dating back to our nation's inception. The Boston Tea Party on December 16, 1773, is a primary example of politically motivated property destruction that provoked a strong reaction from governing officials in Massachusetts and elsewhere (Massachusetts Historical Society, n.d.). As one might expect, states respond in different ways to social movements that challenge existing norms and the social order. In more recent decades, a contingent of the animal liberation and environmental movements have both generated a great deal of attention using extra-legal tactics that include vandalism, theft, arson, and bombing. Consequently, activists have upset a variety of interests (e.g., private corporations, public universities, law enforcement agencies) that have requested the assistance of lawmakers to curtail the tactics that some of them employ.

In 1992, the U.S. Congress passed, and President George H. W. Bush signed, the Animal Enterprise Protection Act (AEPA) to address concerns that industry officials raised about radical animal liberation tactics because they believed that such methods threatened their existence. This change in the law made it a federal crime to damage operations (e.g., farms that raise animals for slaughter) and facilities (e.g., research laboratories, both public and private) that utilize animals to generate profit or knowledge (e.g., advance human health). Then, for more than a decade, radical actions against such enterprises intensified despite this change in the federal code. Underground sister organizations such as the Animal Liberation Front

(ALF) and Earth Liberation Front (ELF) that are each composed of autonomous cells orchestrated coordinated raids at propitious times, that is, often at night under the cover of darkness. In addition, Stop Huntingdon Animal Cruelty (SHAC), which existed as both an organization and campaign, combined above ground (e.g., confrontational protests outside people's homes) and underground tactics (e.g., ALF laboratory raids)—the latter carried out by unknown people on behalf of the cause (Jonas, 2004). These efforts succeeded in causing a considerable amount of disruption and damage to multiple companies and institutions. According to John E. Lewis, deputy assistant director of the Federal Bureau of Investigation's Counterterrorism Division, "the ALF and ELF and related groups have committed more than 1,100 criminal acts in the United States since 1976, and over half of them in the last 8 years resulting in approximately $110 million in damages" (*Animal Rights: Activism vs. Criminality*, 2004). With growing political pressure to address these actions and revisit the AEPA, Congress passed the AETA, which President George W. Bush signed on November 27, 2006.

This study analyzes the testimony and recollections of witnesses from the congressional hearings on extralegal activism, often referred to as "ecoterrorism," that took place between the passage of the AEPA (1992) and AETA (2006). Thirty-five people appeared across six different public hearings; however, only three of them spoke on behalf of the animal liberation and environmental movements, and they each took a different approach. Other witnesses represented such interests as industry, academia, and law enforcement. To better understand the radical activism, elements of cultural theory and legal conscious will be explored through a document analysis of activist testimony and statements prepared for the congressional record as well as their personal reflections on contributing to the democratic process. So, how do an egalitarian bias and an "under the law" legal consciousness manifest in the respective contributions that each of the three activists made during these congressional hearings and reflections upon their participation?

Cultural Theory

Cultural theory rests upon the important foundational work of anthropologist Mary Douglas's (1982) group-grid schema that seeks to categorize people by the social environment they experience. There are several different cosmological types in the schema: (1) low group, low grid; (2) low group, high grid; (3) high group, high grid; and finally; (4) high group, low grid. Here it helps to think of "group" as the X-axis and "grid" as the Y-axis. Group refers to how socially bounded people are to one another. When they are more tightly bounded, that is, high group, they have less freedom to move between different kinds of people besides those who are immediately around them. Grid refers to the number of rules that people are subjected to, that is, how socially prescribed their lives are, as they navigate different interpersonal relationships in different contexts. The higher the grid, the more rules that people have imposed upon them.

Building on the Douglas's work, political scientist Aaron Wildavsky (1987, 1988) uses the schema to develop a theory of culture where propositions are attached to each of the four cultural types in an effort to understand and predict the attitudes and behaviors associated with them. The four types are as follows: individualism (low group, low grid); fatalism (low group, high grid); hierarchy (high group, high grid); and egalitarianism (high group, low grid). One of the multiple ways that you can distinguish between these cultural biases is to examine an issue like human nature because they each have their own distinctive perception of it (Wildavsky et al., 1990). Individualists believe that regardless of the social forces and institutions that exist in a community, people choose their path and shape their destiny. Next, fatalists view the world around them with a great deal of skepticism because they think people are unpredictable, and therefore, they cannot be trusted. Then, hierarchists believe that institutions provide the structure we need to protect ourselves from human behavior that is harmful. Lastly, egalitarians view people as inherently good; however, the institutions of human design lead us astray and undermine our potential.

The existing cultural theory literature that is immediately relevant to this study includes work that focuses on egalitarianism—in general, and animal liberationists and environmentalists—in particular. Wildavsky and Swedlow (1991) find that the egalitarian worldview is more common today than in decades past. The authors gauge this by several indicators, including the number of activists who harbor such sentiments, how strongly the activists believe in their cause, and the number of groups that have formed around issues like animal rights, environmentalism, land use, and feminism. Ellis and Thompson's (1997) analysis of survey data indicates that while most Americans embrace market solutions to environmental problems (e.g., habitat protection), environmentalists who have an egalitarian bias distrust industry and the capitalist system. Ellis (1998) catalogs radical leftist social movements because they exhibit illiberal tendencies when it comes to their respective ideologies and the actions they undertake. One of the groups that he examines is Earth First!, which is more militant in comparison to other mainstream environmental organizations. He notes that not only do these activists distrust the marketplace, but they also espouse "apocalyptic visions" in their critique of our civilization. Duerk (2012) observes that many animal rights activists harbor egalitarian biases, however, most favor utilizing lawful tactics to advance their cause (according to his survey data) and some of them are highly critical of legal institutions (according to his interview data).

Legal Consciousness

For many, the law is a complicated institution—from the language of the code itself to the process that one must navigate if he or she needs to settle a dispute. After conducting 430 interviews, Ewick and Silbey (1998) identified three distinctive types of legal consciousness to understand how people view the legal system. The first of the types is "before the law," and these individuals believe that the law is "objective" and concrete. While they might not always agree with the legal decisions that are rendered, the system is to be respected because of the important role that it plays in society. "With the law" is the second category that Ewick and Silbey find. Here people view the law as a "game" in which the "rules" are subject to change depending on the circumstances. To advance "self-interest," legal actors manipulate the variables that they have control over to produce desired results. The third and last type offered by these authors is "against the law," which means that some people find themselves in conflict with it for their own reasons. As a result, they employ various forms of misbehavior like "omissions," "small deceits," and even "humor" to undermine it as a means of self-preservation rather than exhibit total disdain for the system. These people recognize the law's power; however, they feel compelled to challenge it even if their actions will not change the outcome. Building on Ewick and Silbey's work, Fritsvold (2009) offers yet a fourth type of legal consciousness from his research on radical environmentalists: "under the law." These activists believe that our society has produced a corrupt political system, which extends to a legal system that certain individuals with power use to insulate themselves as well as their interests. Please note the parallel to the cultural theory proposition about egalitarians, more specifically those who champion environmental issues. To destabilize the existing systems (e.g., capitalism), activists engage in "instrumental lawbreaking" to advance their beliefs.

Additional, yet related, studies explore how members of social movements perceive legal institutions, including everything from the law itself to judicial opinions. McCann (1994) finds that wage equity activists use the court system to advance their issue even when they encounter difficulty. They attribute meaning to litigation and must apply different strategies and tactics to overcome the historical discrimination. Furthermore, he explains the "decentered view of law," which contends that our legal system is more than a governing apparatus disconnected from the lives of everyday people. Instead, it is part of a much broader, more complicated, social landscape where political struggle frequently occurs. Here Silverstein's (1996) work is also highly instructive.

She applies the "decentered view of law" to animal liberationists that seek to expand rights language to encompass nonhuman animals because the notion of rights has a different connotation for them when compared to that of people outside the animal liberation movement. It is a matter of how these activists speak about the law as it relates to animals along with the "strategic action" they take to help them.

A Comparison of the AEPA of 1992 and AETA of 2006

Any examination of activist testimony and their subsequent reflections on participating in the democratic process must include a comparison of the two different laws' key provisions, especially since these hearings culminated in elected officials changing the federal code. The Animal Enterprise Protection Act (1992), that is, AEPA, is composed of four major sections: "offense," "aggravated offense," "restitution," and "definitions," while the Animal Enterprise Terrorism Act (2006), that is, AETA, is composed of five major sections: "offense," "penalties," "restitution," "definitions," and "rules of construction." When further examining the verbiage, certain additions become apparent. First, the AETA's "offense" segment extends beyond the animal enterprise itself to safeguard other entities that have either a "connection to" or "relationship with" the animal enterprise. Secondly, AETA addresses activism that "places a person in reasonable fear of . . . death . . . or serious bodily injury" from "a course of conduct involving threats, acts of vandalism, property damage, criminal trespass, harassment, or intimidation," and it protects any targeted person's spouse (or partner) and their children. These changes are a direct response to the aggressive tactics employed by activists associated with the SHAC campaign that focused on multiple companies with business ties to Huntingdon Life Sciences (Best, 2007a,b; Lovitz, 2010). Thirdly, the offense includes not only anyone who "conspires" to violate the statute (a component of both the AEPA and AETA) but also anyone who "attempts" to violate it (which only appears in the AETA). Again, this change appears to be a response to the SHAC activists.

In addition, changes such as those found in the "penalties" and "restitution" segments of the AETA (2006) should be addressed. Whereas the AEPA (1992) only mentions monetary damages in its "offense" section, the AETA does not mention any figures under its "offense" section and focuses solely on this under "penalties" along with more specific punishments for infraction. For example, the AETA states that violating the law could result in "a fine . . . or imprisonment for not more than 5 years, or both, if no bodily injury occurs and . . . the offense results in economic damage exceeding $10,000 but not exceeding $100,000." Sometimes activists employ tactics like vandalism, while other times arson. So, lawmakers built in more structure with the revision to address the varying amounts of damage (and/or harm) that activists cause in pursuit of their goals. They also folded the "aggravated offense" part of the AEPA into the "penalties" segment of the AETA. As for the "restitution" component, provisions exist that address "repeating any experimentation" and "loss of food production or farm income" in both the AEPA and AETA. However, the AETA contains an additional part that reads "a violation . . . may also include restitution . . . for any other economic damage" such as "any losses or costs caused by economic disruption." The latter verbiage is far more expansive in what it might cover regarding an animal enterprise's operation.

Lastly, the "definitions" and "rules of construction" segments merit attention. The AEPA (1992) classifies an animal enterprise as "a zoo, aquarium, circus, rodeo, or lawful competitive animal event," while the AETA (2006) classifies it as "a zoo, aquarium, animal shelter, pet store, breeder, furrier, circus, or rodeo, or other lawful competitive animal event." This extends protection to more businesses that have been the focus of animal liberation activist campaigns. During the late 1990s, fur farms and salons became a popular target for protests and some experienced property destruction (Goodwin, 1998/99). Next, the term "economic damage" has been expanded. The AEPA explains it as "the replacement costs of lost or damaged property or records, the costs of repeating an interrupted or invalidated experiment, or a loss of profits." In contrast, the AETA includes this verbiage and then elaborates with the following: "Increased costs, including losses and increased costs resulting from threats, acts of vandalism, property damage, trespass, harassment, or intimidation taken against a person or entity on account of that person's or entity's connection to, relationship with, or transactions with the animal enterprise." Once again, this language sounds very much like another response to the SHAC campaign's tertiary targeting of companies that had a financial relationship with the biomedical research laboratory (Best, 2007a,b; Lovitz, 2010). It must be noted that while the AETA's verbiage is more extensive, there are questions as to how such "losses" or "costs" would be quantified. After all, "losses" and "costs" attributed to acts of vandalism can be measured more easily than those attributed to trespass or harassment. With regards to the AETA's "rules of construction," officials attempted to ensure that activists who engage in lawful protesting would not have their rights violated. It states that "nothing in this section shall be construed . . . to prohibit any expressive conduct (including peaceful picketing or other peaceful demonstration) protected from legal prohibition by the First Amendment of the Constitution." As of this writing, the AETA has survived a constitutional challenge (*United States of America v. Kevin Johnson and Tyler Lang*, 2015/2017) despite the fact that some think it violates activists' rights by unfairly "targeting" them (Stempel, 2017; Center for Constitutional Rights, 2016).

LITERATURE REVIEW

Since the AETA passage in late 2006, important research has emerged about the politics that spurred legislative action, the provisions of the law itself, and the ramifications of its enactment. Multiple studies claim that dramatic events like the September 11, 2001, attacks heightened concern over activist use of radical tactics that officials now brand "terrorist" acts (Hall, 2006, 2006/2007; Best, 2007a,b; Lovitz, 2010; Salter, 2011; Stanescu, 2014; Loadenthal, 2016; Boyer, 2017). This includes an analysis of the interests directly involved in the lawmaking process and why certain elected officials felt compelled to upgrade the AEPA. Some of the existing work also notes that congressional representatives dismissed the testimony of witnesses like Will Potter who questioned the need for new legislation (Best, 2007a,b; Stanescu, 2014) and how acts of vandalism do not warrant a terrorist label (Stanescu, 2014; Loadenthal, 2016). Among the changes offered by this upgrade, the AETA includes a new provision that addresses tertiary targeting because the SHAC campaign employed it with much success as activists pushed Huntingdon Life Sciences to the edge of financial ruin (Best, 2007a,b; Lovitz, 2010). Another study argues that the AETA should be amended to include whistleblower protection because animal enterprises often do not adequately monitor their practices or embrace the employees, regulators, or activists who point out mistakes or abuses (Hill, 2010). Given some of the law's language, such a protection might be a benefit to activists who often document animal mistreatment. Lastly, different pieces investigate the harm the AETA might cause to the animal rights and environmental movements by discouraging participation in struggles that challenge normative behavior within our social, political, and economic systems (Salter, 2011; Loadenthal, 2016; Boyer, 2017).

METHODOLOGY

This study is a document analysis (Johnson et al., 2020) of witness testimony from, and personal writing about, congressional hearings on extralegal animal liberation and environmental activism in the United States. Such activism is also referred to as "ecoterrorism" by multiple interests—from government to industry officials. While federal lawmakers hosted a total of six hearings on the subject between 1998 and 2006, only three of the thirty-five witnesses who participated represented these social movements. An analysis of the hearing transcripts is necessary to explore how these activists demonstrate an egalitarian bias and an "under the law" legal consciousness. This approach yields a deeper understanding of the activists' worldview and mindset as well as how they differ from one another. The specific hearing transcripts, that is, running records, examined include that of Craig Rosebraugh, Jerry Vlasak, and Will Potter. Craig Rosebraugh, an animal liberation and environmental activist, served as the press officer of the ELF from 1997 until 2001. In addition, he authored a memoir about his experience: *Burning Rage of the Dying Planet: Speaking for the Earth Liberation Front* (2004), which is an important episodic record. Jerry Vlasak is a trauma surgeon and animal liberation activist who currently serves as a press officer for the North American Animal Liberation Press Office (NAALPO). Finally, Will Potter is a journalist, animal liberation activist, publisher of GreenIsTheNewRed.com, and author of *Green Is the New Red: An Insider's Account of a Social Movement under Siege* (2011)—the last of which is another vital episodic record that must be examined. Both Rosebraugh's and Potter's respective books are studied here because they offer reflections that are immediately relevant to this research.

ANALYSIS OF TRANSCRIPTS AND MEMOIRS

Craig Rosebraugh: Contumacious Environmentalist

Several months after the September 11 terrorist attacks, the Subcommittee on Forests and Forest Health convened a hearing entitled, "Eco-terrorism and Lawlessness on the National Forests" (2002). House members subpoenaed Craig Rosebraugh to appear and answer a series of questions because the ELF had committed multiple extralegal acts and he had served as the group's press spokesperson from 1997 to 2001. Having been a longtime activist, he understood the political gravity of the situation. As evidenced by the transcript and his memoir, he demonstrated an egalitarian worldview as well as an "under the law" (Fritsvold, 2009) legal consciousness prior to his participation, during his appearance before lawmakers, and after the hearing ended. Here it must be noted that Fritsvold (2009) analyzes some of Rosebraugh's views in his own study; however, this research explores more of Rosebraugh's ideas and actions.

In *Burning Rage of a Dying Planet*, Rosebraugh recalled his plan to resist congressional officials' request that he participate on February 12 by eluding them. Such recalcitrance had become a habit whenever the government issued him a subpoena:

> I vowed to delay receiving it as long as I could . . . this sort of thing had happened seven times before, I had survived each time without cooperating and had never been indicted with single crime related to my spokesperson work,

> and since it was the House I would only risk one year in jail for not cooperating—I decided to carry on with my regular life as much as possible, while trying to make those serving the subpoena earn their pay. (Rosebraugh, 2004)

For weeks, officers from the U.S. Marshals Service attempted to serve him at his residence, but he managed to evade them on multiple occasions—sometimes by simply not answering the door, another time by fleeing in a vehicle onto the nearby highway. He and the others living in the house even covered the windows on the first floor so that no one could tell if people were home. Eventually, the Marshals caught him by surprise and served him when he stepped out onto the porch one evening (Rosebraugh, 2004).

Subsequently, the day that Rosebraugh testified before the subcommittee, he refused to answer most of the questions that members asked. On fifty-four separate occasions, he invoked the Fifth Amendment. His approach irritated some of those in attendance. Representative Inslee (D) of Washington claimed that Rosebraugh could not legally do this because he had already submitted a statement (*Eco-terrorism and Lawlessness on the National Forests*, 2002). Moreover, Representative McInnis (R) of Colorado threatened incarceration for noncompliance (*Eco-terrorism and Lawlessness on the National Forests*, 2002). Neither ploy induced cooperation with their inquest. . So, members later sent a list of fifty-four follow-up questions for him to answer. As one might expect, he responded with "I don't know" and "I do not recall" to many items (*Eco-terrorism and Lawlessness on the National Forests*, 2002). However, he responded to some questions, including the following:

> Do you find it disconcerting that, when ELF firebombed forestry research labs at the University of Washington and Minnesota in 2001 and 2002 respectively, the fire quickly spread to other areas on both campuses, potentially endangering lives in buildings not targeted by ELF? In the case of the University of Washington, the fire spread to an adjacent library. And in the case of the University of Minnesota, the man-made fire spread to the soils testing center in the near vicinity. (*Eco-terrorism and Lawlessness on the National Forests*, 2002)

To which Rosebraugh replied:

> I do not find it disconcerting that the ELF firebombed, without physically harming anyone, research into genetic modification of our natural world for profit. Genetic engineering is a threat to life on this planet. As for other factual allegations, I do not know whether or not they are true, so I do not feel comfortable commenting on them. (*Eco-terrorism and Lawlessness on the National Forests*, 2002)

This is significant for multiple reasons. First, while Rosebraugh's level of obstinacy did not exactly match that of his appearance before the subcommittee, he openly defended the ELF despite having stepped down as its spokesperson. In short, he refused to disavow the group's extralegal tactics. Secondly, he critiqued the industry. To him, genetic engineering is a destructive capitalist endeavor worthy of condemnation. Finally, he refused to acknowledge reports that the fires caused damage to any other facilities besides those targeted by the ELF activists.

While Rosebraugh's testimony, or lack thereof, served as indicators of his egalitarian worldview, the lengthy prepared statement that he submitted in advance goes much further. He felt that the circumstance implored him to make an unadulterated written contribution to the congressional record:

> I knew for certain that I was not going to provide any verbal testimony, but I thought that entering something in writing might give me the opportunity to voice my opinion against the state's terrorism and its support of environmental destruction. Additionally, as I was the only pro-environmental representative included in the hearing, I felt some responsibility to at least make an objection to the hearing and take a stand in support of the ELF and ALF. (Rosebraugh, 2004)

The most revealing segments of this reflection from his memoir are the reference to "state's terrorism" and "take a stand in support of the ELF and ALF" because the former is a harsh judgment of policy and the latter condones activists who employ extralegal tactics to achieve their goals. Shifting to the statement itself, Rosebraugh wrote the following:

> Tactics . . . approved by the U.S. government . . . rarely . . . challenge or positively change the very entities that are responsible for oppression . . . In the mid-1990s, individuals angry and disillusioned with the failing efforts to protect the natural environment through state sanctioned means, began taking illegal action . . . Immediately, the label of ecoterrorism appeared in news stories describing the actions of the Earth Liberation Front. Where exactly this label originated is open for debate, but all indications point to the federal government of the United States in coordination with industry and sympathetic mass media. (*Ecoterrorism and Lawlessness on the National Forests*, 2002)

Here he claimed that lawful approaches to social change often fail to deliver just outcomes because those in power have the capacity to undermine activist efforts. Furthermore, he attributed the origin of the "ecoterrorism" label to the public and private interests that are most threatened by those who employ radical tactics. Ultimately, the label is a negative, political stigma assigned to people who take risks to protect animals and the environment.

The media is invoked as sharing responsibility for disseminating the ideas that disparage those who refuse to adhere to normative political channels to champion these causes. The implication is that activists must resort to extralegal methods because a corrupt system makes ecological degradation possible.

Jerry Vlasak: Didactic Animal Liberationist

On October 26, 2005, the U.S. Senate Committee on Environment and Public Works hosted a hearing entitled, "Eco-terrorism," where they invited Jerry Vlasak, M.D., to speak on behalf of the NAALPO. During his testimony he explained why activists resort to radical tactics, thereby demonstrating an egalitarian bias and an "under the law" legal consciousness. In his statement to lawmakers, Vlasak critiqued Huntingdon Life Sciences (HLS)—a research laboratory that had been the focus of animal liberation activists for several years:

> Huntingdon Life Sciences kills 500 animals a day. . . . They carry out experiments which involve poisoning and torturing animals to death with household products, pesticides, drugs, herbicides, food colorings, sweeteners, oven cleaner, and cosmetics . . . They have been infiltrated and exposed in undercover investigations five times in recent years by journalists and animal rights campaigners. Each time horrific evidence of animal abuse and staff incompetence has been uncovered, including workers punching beagle puppies in the face, simulating sex with animals in their care, dissecting primates while they are still alive and falsifying experiments to get their client's product on the market. (*Eco-terrorism*, 2005)

Then, he expanded his scope to more broadly address the business of animal research along with specific models, that is, both events and figures, which members of the animal liberation movement draw inspiration from when challenging institutions that they believe perpetuate injustice:

> Huntingdon is the posterchild of an abhorrent, unnecessary and wasteful industry that not only murders millions of innocent, suffering animals, but dooms countless humans to their own unnecessary suffering because scarce healthcare dollars are wasted on useless animal research and testing. The struggle for animal liberation needs to be seen in historical context, like the Boston Tea Party ignited a revolution, like Nelson Mandela and his fight against apartheid, like the suffragettes and John Brown . . . these noble and historical figures fought the governmental powers of oppression, slavery, and exploitation. (*Eco-terrorism*, 2005)

These models represent parallels. He viewed animal liberation as another manifestation of resistance to institutionalized power that undermines human and nonhuman animal life.

In addition, as a press officer who disseminates information about extra-legal activism, Vlasak (like Rosebraugh) questioned the media's ability to cover such stories. He accused them of shirking their responsibility—at best, and misrepresentation—at worst, while automatically defaulting to the side of businesses that mistreat animals as part of their operation:

> The actions of activists who care enough about animals to speak out . . . and . . . risk their own lives and freedom have a message that is most urgent and one that deserves to be heard and understood. Often, acts of animal liberation either go unreported in the media or are uncritically vilified as violent and terrorist, with no attention paid to the suffering the industries and individuals gratuitously inflict upon animals. (*Eco-terrorism*, 2005)

To him, media outlets never inquired why activists resort to radical tactics, and consequently, a vital component of the story is missing, which is a detriment to the public interest. More importantly, needless animal suffering continues without being questioned or challenged. In other words, the media demonstrated complicity in his mind.

During the hearing, Senator Inhofe (R) of Oklahoma raised the issue of the necessity of using animal subjects in medical research. Here he asked Vlasak: "Do your fellow animal rights activists understand that animal testing is required by law and therefore, the people who are performing this testing are merely following the law?" (*Eco-terrorism*, 2005). Vlasak responded by making a connection between how people and animals have experienced legally justified ill-treatment. He said:

> I understand that they are merely following the law, and the law in this case is wrong, just like the law allowed slavery was wrong at one time . . . Non-human lives, non-human animal lives, are as precious as human lives. At one time, racism and sexism and homophobism were prominent in our society. Today speciesism is prominent in our society. It is just as wrong as racism. (*Eco-terrorism*, 2005)

This answer demonstrated an "under the law" legal consciousness. First, it addressed the legally disadvantaged position that nonhuman animals currently experience by drawing a parallel to the lesser position that African Americas, women, and homosexuals have historically experienced. Secondly, it attributed a value to nonhuman animal life that mainstream society has yet to recognize as people adhere to a legal system that he believed to be deeply flawed.

Will Potter: Oppositional Journalist

Independent journalist Will Potter addressed the Subcommittee on Crime, Terrorism, and Homeland Security on May 23, 2006, after members invited him to speak about H.R. 4239, that is, the AETA. As mentioned, the AETA revised the AEPA by adding additional sections and rewriting or editing some others. Consistent with an egalitarian bias and an "under the law" legal consciousness, Potter challenged officials about the necessity and implications of the proposed legislation. His criticism of governing institutions, including law enforcement, and the marketplace, that is, industry, is apparent early in his testimony:

> I've closely followed the animal rights and the environmental movements, and the corporate-led backlash against them. I've documented an increasingly disturbing trend of terrorist rhetoric, sweeping legislation, grand jury witch hunts, blacklists, and FBI harassment reminiscent of tactics used against Americans during the Red Scare. The Animal Enterprise Terrorism Act is a continuation of that trend. (*Animal Enterprise Terrorism Act*, 2006)

When Potter used a phrase like "corporate-led backlash," one can infer that he spoke of the results that business officials have achieved in their appeals to congressional members. Their political efforts have manifested in a variety of different ways—all of which undermine the animal liberation and environmental movements. To him, they have wrongly framed the extralegal activism as terrorist activity to further insulate companies from those who threaten their existence.

Furthermore, Potter contested the need for a new legislation. Initially, he said that criminal statutes already protect these industries. Then, he asserted that common forms of activism would be criminalized and perhaps categorized as terrorism because they could be employed to financially undermine companies that activists choose to focus on:

> Law enforcement has not proven the need for heavier-handed tactics. Property crimes are already punishable as so-called animal enterprise terrorism. This bill, though, further expands that sweeping category to include protests, boycotts, undercover investigations, whistleblowing, and non-violent civil disobedience. The bill criminalizes any activity against an animal enterprise or any company tangential to an animal enterprise that causes economic damage . . . including the loss of profits. That's not terrorism, that's effective activism . . . Businesses exist to make a profit, and if activists want change, they have no choice but to tug at the purse strings. (*Animal Enterprise Terrorism Act*, 2006)

Here he also mentioned the entities that have a financial relationship with the animal enterprise itself. This is one of the major enhancements that the

AETA offered. The SHAC campaign achieved great success through its utilization of tertiary targeting, which prompted many companies to sever ties with Huntingdon Life Sciences thereby causing HLS immense financial distress.

Social movements are never ideologically monolithic and most animal liberation and environmental activists do not violate laws to advance their specific issues. Subsequently, Potter conveyed uneasiness that the AETA would discourage lawful advocacy because people were afraid of what harm the ecoterrorist label might do to them:

> Perhaps the greatest danger of this legislation, though, is that it will impact all animal activists, even those that never have to enter a courtroom. The reckless use of the word "ecoterrorism" by corporations and the government has already had a chilling effect, and this legislation will compound it. Through my reporting I've already heard the widespread fears of activists that they may soon be labeled terrorists, even for legal activity. (*Animal Enterprise Terrorism Act*, 2006)

One of the major implications and ironies here is that inducing a "chilling effect" with a label and legislation hinders participation in a political system that needs such engagement to function. Whether perceived or real, activists communicated their fears to him, which means they felt their rights were threatened by government encroachment.

Five years later when reflecting upon his experience, Potter questioned the whole democratic process—a process that he once believed he could make a meaningful contribution to through his appearance before the subcommittee. His egalitarian critique more thoroughly expanded to include not only the legislation and industry but more broadly the political system:

> I had naively thought that lawmakers had invited me to listen . . . ask questions, and then come to their own conclusions. . . . Republicans and Democrats alike had already been swayed by more influential players. . . . I was . . . invited to appear as a token gesture of dissent in their spectacle of democracy. . . . Corporations had been awaiting their moment for years and wanted nothing more than for their bill to proceed unchallenged. (Potter, 2011)

Referring to the invitation he received, the phrase "token gesture of dissent in their spectacle of democracy" suggests that those who arranged the hearing did not have any intention of listening to him because some political actors had already determined the endpoint: the passage of the AETA. Members of Congress dismissed his concerns about it while favoring business interests that successfully manipulated the system to their advantage.

Like Rosebraugh and Vlasak, Potter next questioned the integrity of those in his chosen craft. He claimed that there is an inherent bias in how reporters approach covering news stories: "Too often, journalists report 'both sides' as if they are equal, even when one side is riddled with lies and motivated by self-interest. Over-reliance on 'official' government and corporate sources, and deference to their word, has taken precedence over critical, investigative reporting that speaks truth to power" (Potter, 2011). Not only did he critique our political system, but also the industry that disseminates information to us as citizens and consumers. If reporters favor one source over another, then the truth might be lost when it comes to important issues—especially issues that involve the environment and animals because certain interests have strong incentives to withhold details.

CONCLUSION

The continued employment of extralegal tactics by radical contingents of the animal liberation and environmental movements during the 1990s and 2000s compelled members of Congress to convene six public hearings, which eventually produced the AETA in 2006. Of the thirty-five witnesses who participated, only three of these spoke on behalf of these social movements. This study aimed to answer the following question: How does an egalitarian bias and an "under the law" legal consciousness manifest in the respective contributions that each of the three activists made during the congressional hearings and reflections upon participation in them?

An analysis of Craig Rosebraugh, Jerry Vlasak, and Will Potter's testimony, statements, and memoirs yields a finding of similarities that demonstrate an egalitarian bias and an "under the law" legal consciousness as well as certain differences in the way that they each approached their participation within the political system. They not only criticized government and industry but also accused them of collusion to the detriment of animals, the environment, the democratic process, and public. Furthermore, they espoused a distrust of media's ability to accurately report on animal liberation and environmental activism, which they argued only disseminates misinformation.

Now, where they differ is the way in which they addressed federal lawmakers. Craig Rosebraugh refused to fully cooperate with multiple government officials before, during, and after his appearance in Washington, D.C. This appears to be, at least in part, a result of the fact that lawmakers subpoenaed him to testify before them. Much of his behavior is that of contumacious activist. In contrast, Jerry Vlasak used the hearing as a platform to illuminate the plight of animals in places like research laboratories. Moreover, he drew parallels between animal liberation and other political struggles in history that have challenged people to rethink the social order that has been constructed and reinforced. Officials invited him to address the committee, but he had the choice to decline as they did not issue him a subpoena. This makes his participation a didactic endeavor. Lastly, Will Potter accepted an invitation to appear before the congressional subcommittee where he not only questioned the need for new legislation but also expressed his trepidation about classifying some forms of activism as terrorism because that could discourage people from partaking in lawful efforts to improve the condition of animals. He has been labeled oppositional here for both the challenge he offered as well as the concerns that he shared with subcommittee members.

REFERENCES

Animal Enterprise Protection Act of 1992, Pub. L. No. 102-346, 106 Stat. 928 (1992).

Animal Enterprise Terrorism Act, Subcommittee on Crime, Terrorism, and Homeland Security of the Committee on the Judiciary, 109th Cong. (2006) (testimony of Will Potter).

Animal Enterprise Terrorism Act of 2006, Pub. L. No. 109-374, 120 Stat. 2652 (2006).

Animal Rights: Activism vs. Criminality, Hearing before the Committee on the Judiciary, 108th Cong. (2004) (testimony of John E. Lewis). https://www.govinfo.gov/content/pkg/CHRG-108shrg98179/html/CHRG-108shrg98179.htm.

Best, S. (2007a). Dispatches from a police state: Animal rights in the crosshairs of state repression. *The International Journal of Inclusive Democracy, 3*(1). https://www.inclusivedemocracy.org/journal/vol3/vol3_no1_Best_animal_rights.htm.

Best, S. (2007b). The Animal Enterprise Terrorism Act: New, improved, and ACLU approved. *Journal for Critical Animal Studies, 5*(1), 66–88.

Boyer, C. (2017). Examining the extent and impact of surveillance on animal rights activists [Unpublished master's thesis]. University of Nevada, Las Vegas.

Center for Constitutional Rights. (2016, September 21). "Terrorism" law targeting animal rights activists is unconstitutional, attorneys argue in court [Press release]. https://ccrjustice.org/home/press-center/press-releases/terrorism-law-targeting-animal-rights-activists-unconstitutional.

Douglas, M. (1982). *In the active voice*. Routledge & Kegan Paul.

Duerk, J. (2012). A cultural analysis of the animal rights movement [Unpublished doctoral Dissertation]. Northern Illinois University.

Eco-terrorism and Lawlessness in the National Forests: Oversight Hearing before the Subcommittee on Forests and Forest Health of the Committee on Resources, 107th Cong. (2002). https://www.govinfo.gov/content/pkg/CHRG-107hhrg77615/html/CHRG-107hhrg77615.htm.

Eco-terrorism and Lawlessness in the National Forests, Oversight Hearing before the Subcommittee on Forests and Forest Health of the Committee on Resources, 107th

Cong. (2002) (testimony of Craig Rosebraugh). https://www.govinfo.gov/content/pkg/CHRG-107hhrg77615/html/CHRG-107hhrg77615.htm.

Eco-terrorism: U.S. Senate Committee on Environment and Public Works, 109th Cong. (2005). https://www.epw.senate.gov/public/index.cfm/2005/10/full-committee-hearingeco-terrorism.

Eco-terrorism, U.S. Senate Committee on Environment and Public Works, 109th Cong. (2005) (testimony of Jerry Vlasak). https://www.epw.senate.gov/public/index.cfm/2005/10/full-committee-hearingeco-terrorism.

Ellis, R. J. (1998). *The dark side of the left: Illiberal egalitarianism in America.* University of Kansas Press.

Ellis, R. J., & Thompson, F. (1997). Culture and the environment in the Pacific Northwest. *American Political Science Review, 91*(4), 885–897.

Ewick, P., & Silbey, S. (1998). *The common place of law: Stories from everyday life.* University of Chicago Press.

Fritsvold, E. D. (2009, Fall). Under the law: Legal consciousness and radical environmental activism. *Law & Social Inquiry, 34*(4), 799–824.

Goodwin, J. P. (1998/99, Winter). Fur wars '98. *No Compromise*, 11, 1, 4. http://thetalonconspiracy.com/2011/11/no-compromise-9-10-11/.

Hall, L. (2006, October 31). The Animal Enterprise Terrorism Act may soon be law: How could this happen? *Dissident Voice.* http://www.dissidentvoice.org/Oct06/Hall31.htm.

Hall, L. (2006/2007, December/January). Working for the clampdown: How the Animal Enterprise Terrorism Act became law. *Satya.* http://www.satyamag.com/dec06/hall.html.

Hill, M. (2010). The Animal Enterprise Terrorism Act: The need for a whistleblower exception. *Case Western Reserve Law Review, 61*(2), 649–678.

Jonas, K. (2004). Bricks and bullhorns. In S. Best and A. J. Nocella (Eds.) *Terrorists or freedom fighters? Reflections on the liberation of animals* (pp. 263–271). Lantern Books.

Loadenthal, M. (2016). Activism, terrorism, and social movements: The "green scare" as monarchical power. *Social Movements, Conflicts and Change, 40*, 189–226.

Lovitz, D. (2010). *Muzzling a movement: The effects of anti-terrorism law, money, and politics on animal activism.* Lantern Books.

Massachusetts Historical Society. (n.d.). The Boston Tea Party. https://www.masshist.org/revolution/teaparty.php.

McCann, M. W. (1994). *Rights at work: Pay equity reform and politics of legal mobilization.* The University of Chicago Press.

Potter, W. (2011). *Green is the new red: An insider's account of the social movement under siege.* City Lights Books.

Rosebraugh, C. (2004). *Burning rage of a dying planet: Speaking for the Earth Liberation Fron*t. Lantern Books.

Salter, C. (2011). Activism as terrorism: The green scare, radical environmentalism, and governmentality. *Anarchist Developments in Cultural Studies, 1*, 211–238.

Silverstein, H. (1996). *Unleashing rights: Law, meaning, and animal rights movement.* The University of Michigan Press.

Stanescu, V. (2014). Kangaroo court: Analyzing the 2006 "Hearing" on the AETA. In J. Del Gandio and A. Nocella II (Eds.) *The terrorization of dissent: Corporate repression, legal corruption, and the Animal Enterprise Terrorism Act* (pp. 51–67). Lantern Books.

Stempel, J. (2017, November 8). Animal activists who freed 2,000 minks lose U.S. appeal. *Reuters*. https://www.reuters.com/article/us-usa-animal-rights-minks/animal-activists-who-freed-2000-minks-lose-u-s-appeal-idUSKBN1D82Q1.

United States of America v. Kevin Johnson and Tyler Lang, 14-CR-390 (D. Ill. 2015), aff'd, 16-1459 and 16-1694 (7th Cir. 2017).

Wildavsky, A. (1987, March). Choosing preferences by constructing institutions: A cultural theory of preference formation. *American Political Science Review, 81*(1), 3–21.

Wildavsky, A. (1988, June). Political culture and political preferences. *American Political Science Review, 82*(2), 589–597.

Wildavsky, A., & Swedlow, B. (1991). Is radical egalitarianism really on the rise? In A. Wildavsky (Ed.) *The rise of radical egalitarianism* (pp. 63–98). The American University Press.

Wildavsky, A., Thompson, M., & Ellis, E. (1990). *Cultural Theory*. Westview Press.

Part III

POLICY

Chapter 8

Changing Lanes and Changing Places

An Examination of Race, Urban Bikeways, and Gentrification in American Cities

Markie McBrayer

INTRODUCTION

While the environmental justice literature establishes that environmental burdens are often concentrated among poor communities of color, little scholarship assesses the distribution of *positive* sustainability policies and how those policies, in turn, reshape those communities. Only recently have scholars started examining how green policies are distributed within a city (Reames, 2016; Rigolon et al., 2018) and how those policies might gentrify a community (Lubitown, 2016; Stehlin, 2015).

Here, I explore how a unique sustainability policy—bicycle infrastructure—is distributed throughout nine urban areas. Based on the environmental justice literature, I expect that communities of color will lack access to bicycle infrastructure due to the fact that sustainable infrastructure potentially reinforces preexisting racial and ethnic hierarchies and inequities. However, findings demonstrate that minority communities have greater relative access to bicycle infrastructure than White communities. Second, I examine the association between bike lanes and gentrification—gentrification being conceptualized as inflows of middle and upper classes into traditionally low-income neighborhoods (Glass, 1964). Using Wyly and Hammel's (2013) operationalization of gentrification, I find that some facets of gentrification—communities where the population grew, communities where the percent of employed people grew, communities where the college-educated population grew, and communities where the median household income grew—are associated with greater access to bicycle infrastructure. Ultimately, this piece explores who has access to cycling infrastructure, thereby shining light on municipal service provision, or how services and policies are distributed in urban areas. Moreover, this work assesses how investments in bikeways might encourage gentrification by attracting more affluent residents.

THE DISTRIBUTION OF ENVIRONMENTAL BURDENS AND BENEFITS

In seminal pieces, scholars and researchers demonstrate that minority communities bear a greater environmental burden than their White counterparts (Bullard, 1983; Government Accountability Office, 1983; United Church of Christ, 1987). More recent research has come to similar conclusions. In 2007, Bullard, Mohai, Saha, and Wright find that neighborhoods with hazardous waste facilities have populations where 56 percent are minorities, while neighborhoods without hazardous waste facilities have minority groups comprising approximately 30 percent of the population. Wolverton (2009) finds that waste sites are more likely to be sited in communities of color. Beyond hazardous waste sites, Depro and Timmins (2012) find that poor and minority households have greater ozone exposure. The reason that environmental burdens tend to be clustered in communities of color is less clear.

In terms of why toxic waste facilities and pollutants seem to be concentrated in low-income and minority communities, there are two possible causal explanations (Mohai et al., 2009). First, minority communities lack the political power that would discourage environmental burdens being sited in their community (Bullard, 2018; Hamilton, 1995). Governments and firms are cognizant of this lack of political power among vulnerable communities, and thus, capitalize on this when siting waste facilities (Cole & Foster, 2001). Second, businesses and firms are not intentionally trying to discriminate against minorities. Instead, they are simply trying to maximize profits, and in turn site these facilities in places where labor and land are both inexpensive, which is oftentimes in poor, minority communities (Ringquist, 2000; Roberts, 2001).

Nevertheless, there is some disagreement about the temporal ordering of events. Specifically, are environmental hazards concentrated in communities of color? Or are low-income individuals and minorities more likely to move into places with greater proportions of environmental burdens. In one camp, Crowder and Downey (2010) find that Black and Latino households are more likely to move into communities with greater environmental burdens than their White counterparts. Conversely, in examining the distribution of polluting facilities over decades-long time-spans, Pastor et al. (2001) and Saha and Mohai (2005) find that hazardous waste facilities are sited in areas with greater proportions of minority and low-income residents, rather than shifting mobility patterns. In other words, hazardous waste disposal facilities are sited *first* in vulnerable communities, as opposed to vulnerable communities moving into these locations with higher concentrations of environmental burdens. Nevertheless, there's still very little research on how positive sustainability policies intended to improve or protect a city's biophysical environment (e.g., local greenhouse gas reductions, parks, investments in alternative transit), are distributed throughout a city, and whether vulnerable communities have access to these amenities. The research herein assesses where bikeways are sited in urban areas and which demographic groups have greatest access to them.

Research on the distribution of positive sustainability policies is decidedly mixed, with most work exploring the distribution of parks, recreation facilities, public transit, and highway infrastructure. In examinations of parks and recreation facilities, some work concludes that poorer communities with greater proportions of minorities are considered to have improved access, while other work finds the opposite. For instance, in their examination of how a city's demographics affect the quality of its parks, Rigolon et al. (2018) find that cities with higher incomes and lower proportions of minorities have higher quality parks than those with greater proportions of low-income residents and minorities. Indeed, most research shows that wealthier and whiter communities have superior park quality (Jenkins et al., 2015; Vaughan et al., 2013). This is further confirmed in other work analyzing neighborhood-level data, where higher and middle SES individuals have greater access to parks (Rigolon & Flohr, 2014; Willemse, 2013). Yet, other research demonstrates the opposite. For instance, Comer and Skraastad-Jurney (2008) and Jones and Coombes (2009) show that individuals from lower socioeconomic brackets live somewhat closer to parks. Ultimately, these mixed results are likely a function of how policy distribution—that is, who receives access to quality services and amenities—is measured, whether it is access to quality parks or if it is more a function of distance or acreage (Rigolon, 2016).

Transportation justice provides a similar lens when examining which individuals and communities have greatest access to positive environmental policies, like public transport and highways. Public transit is of course rooted in a history of segregation, with there being separate rail, trolley, and buses for people of color after *Plessy v. Ferguson (1896).* While these explicitly discriminatory policies are prohibited now, inequities persist. For instance, the fare revenue from short trips that are often taken by low-income passengers typically subsidizes the longer commuter rail fares that are taken by wealthier suburban passengers (Taylor et al., 1995; Taylor & Morris, 2015). These fare disparities are of course exacerbated by the fact that federal funding for transportation projects has been proportionally split since the 1970s, where

80 percent is devoted to highway projects, which primarily benefits middle- and upper-class suburban dwellers, while public transportation receives only 20 percent (Nall, 2018). Most of the work examining transportation justice (Delbosc & Currie, 2011; Litman, 2015; Kramer & Goldstein, 2015; Sanchez, 2008) focuses on public transportation and highway infrastructure, with far less scholarship examining bike lanes.

BIKE LANES, SUSTAINABILITY, AND GENTRIFICATION

Urban and environmental planners have increasingly advocated for bike lanes as a way to reduce greenhouse gases and improve air quality. Neves and Brand (2019) find that if short car trips are substituted with walking and biking, it would reduce CO_2 emissions by approximately 5 percent overall. At the individual level, access to bicycles is associated with lower CO_2 emissions (Brand et al., 2013). In fact, Blondel et al. (2016) estimate that greenhouse gas (GHG) emissions are more than ten times lower than the GHG emissions associated with a motor vehicle. In terms of air quality, Johansson et al. (2017) estimate in Stockholm County that if all commuters with less than a 30-minute bicycle commute opted for bicycle as opposed to motor vehicle, it would reduce NO_x and black carbon exposure, and it would save approximately 450 lives per year in a county of more than 2 million. Given these environmental benefits, many residents and practitioners advocate for infrastructure to support bicycling.

Traditionally bike-planning is seen as being a largely technocratic endeavor that is simply centered on the provision of a relatively minor piece of infrastructure (Lubitow & Miller, 2013; Lindelöw et al., 2016; Lubitow et al., 2016; Zavestoski & Agyeman, 2015). Yet, Goodman (2010) notes that the planning and implementation of bicycle infrastructure can evoke "unusual scenes of friction" (A26). Some of this disagreement is due to a reduction in the number of parking spots or road space for vehicles (Spotswood et al., 2015; Vivanco, 2013). Shaer (2011) and Stein (2011) both note though that bike lanes are ripe as spaces for cultural and class conflict. Given the rapidly changing demographics of urban centers, this is no surprise (Greenblatt, 2018). Moreover, the champions of bike lane infrastructure tend to be perceived as mostly among the young, White creative class (Florida, 2005). And this stereotype is not unwarranted. In fact, many cities across the United States have explicitly stated that investments in bike lanes is intended to attract the creative class (Walljasper, 2012). The primary concern is whether investments in bike lanes, in turn, result in gentrification.

Ruth Glass, the original scholar to neologize the term gentrification, explains the process of gentrification, which, according to her, is the process where lower class neighborhoods are slowly transformed by newer middle-class residents:

> One by one, many of the working class quarters of London have been invaded by the middle classes—upper and lower. Shabby, modest mews and cottages—two rooms up and two down—have been taken over, when their leases have expired, and have become elegant, expensive residences. . . . Once this process of "gentrification" starts in a district, it goes on rapidly until all or most of the original working class occupiers are displaced, and the whole social character of the district is changed. (1964, p. xviii)

And these "middle classes" might be attracted to investments in amenities, like bicycle infrastructure. The concern then is whether governments' investment in bicycle infrastructure can lead to inflows of these middle classes, thus resulting in the transformation and displacement of original, often low-income residents (Checker, 2015).

And indeed, a growing strand of literature has begun to examine how investments in bike lanes or changes in commuting patterns might encourage gentrification. Specifically, research and commentary points to the relationship where gentrification encourages investments in public services. DeSena describes how recent inflows of higher classes change the public services in a Brooklyn neighborhood: "New gentrifiers expect accommodation for their lifestyle, and seek to impose their desired lifestyle upon the existing community. New gentrifiers do not look to 'fit in' to Greenpoint, but instead, they attempt to change it to meet their own needs while disregarding the local history and culture" (2012). In a public interview, John Stehlin, a geographer at the University of California-Berkeley, echoes a similar sentiment: "Cycling feeds into wider urban changes, including gentrification, but it does not *cause* gentrification. A bicycle lane gets put on a street that is already undergoing change" (Geoghegan, 2016). Herrington and Dann (2016) come to a similar conclusion when they examine gentrification and bicycle use in Portland. Specifically, they find that communities that have grown substantially whiter and more educated are more likely to see vast increases in bicycle use. Hoffman also argues that "where gentrification goes, bicycle infrastructure follows quickly behind" (2016). Still, she acknowledges that bike lanes might be used by people and groups in power to "erase communities" (2016). Indeed, Flanagan et al. (2017) find that greater investments in bicycle infrastructure are associated with both communities that are more privileged or have become more privileged over the 1990–2010 period. Still, these analyses are limited to either case studies or statistical analyses of two cities, thereby limiting the generalizability of their results.

THEORETICAL EXPECTATIONS

I expect that communities of color will have less access to bike lanes than communities with higher proportions of non-Hispanic Whites. I posit that this is because of the aforementioned causal mechanisms outlined in the environmental justice literature. Specifically, communities with higher proportions of minorities might lack the political power to advocate for the infrastructure. In contrast, communities with higher proportions of non-Hispanic Whites will be able to demand and, in turn, receive the infrastructure that they want because of their greater political power. As a consequence, I posit the following hypothesis:

Hypothesis 1: Communities with greater proportions of non-Whites will have less access to bike lanes than communities with greater proportions of non-Hispanic Whites.

At the same time, I expect that the presence of bike lanes is associated, in some ways, with gentrification. As previous scholars have outlined, gentrification brings new residents in who have different transportation preferences, like bicycles, than original residents. Since these new residents tend to have greater political power than preexisting residents, inflows of middle- and upper-class individuals are able to successfully demand new bicycle infrastructure. Alternatively, investments in bicycle infrastructure in low-income communities might attract middle- and upper-class individuals who desire to live in a more bikeable community, as DeSena (2012) suggested. Ultimately, these two potential causal mechanisms are associated with the following second hypothesis:

Hypothesis 2: Communities with greater access to bike lanes will be more likely to experience gentrification than communities with less access to bike lanes.

It is important to note that I do not explore whether gentrification causes investments in bike lanes or whether investments in bike lanes cause gentrification. Rather I examine whether there is an association between these two phenomena.

DATA AND METHODS

In order to assess the spatial distribution of bikeways, I gathered publicly available shapefiles—a common file that stores geographic information that can then be used in mapping software, like ArcGIS—of the bicycle infrastructures for nine of the ten most populous cities in the United States. This spans six states and includes Chicago, Dallas, Houston, Los Angeles, New York, Philadelphia, Phoenix, San Diego, and San Juan. This excludes San Antonio because of data availability and inconsistency in their publicly available shapefiles. These shapefiles are an attempt to gather the most recent form of bicycle infrastructure in these nine cities.

In ArcGIS (a common mapping software), these shapefiles are mapped and overlaid with shapefiles of Census 2010 block groups. Block groups are typically an aggregation between five and ten blocks, and they serve as somewhat of a proxy for neighborhoods and communities and are commonly employed in neighborhood-level analyses (Alaniz et al., 1998; Alaimo et al., 2010). After mapping the centroid—that is, the geographic center of these spatial units—of these block groups, I determined the distance between these centroids and the nearest bike lane.

Because each of these cities have very different leve1 1ls of density, sprawl, and total area, analyses attempt to control for differences among cities. For instance, New York City has approximately 300 square miles of land, while Houston has nearly 700 square miles of land. Moreover, New York City has 28,211 residents per square mile, while Phoenix has 3,126 people per square mile.[1] As a consequence, I standardize the distances within these cities. Specifically, I measure each city's mean distance between their block groups and bike lane. Then I subtract that mean distance from each block group's distance to their nearest bike lane and then divide by that specific city's standard deviation. This allows us to more easily compare across cities.

For the racial and ethnic composition measures, I employ five-year ACS measures from 2018 for block groups. Specifically, I measure the percent of non-Hispanic White, percent Hispanic, percent Black, and percent non-White. Much like the distance measures, I standardize these racial and ethnic measures, given the varying racial and ethnic compositions of these cities. For instance, in Phoenix, approximately 66 percent of the population is non-Hispanic White, while New York City's non-Hispanic White demographic comprises 43 percent. This helps to better account for the variation in demographic composition across cities.

To test the relationship between gentrification and bike lanes, I employ Hammel and Wyly's (1996) Census measures of gentrification. For my purposes, I use the following of their measures:

- Median household income,
- Change in median household income,
- Percentage of persons age twenty-five and over with a college degree or more,

- Change in percentage of persons age twenty-five and over with a college degree or more,
- Percentage of employed people,
- Change in percentage of employed people,
- Median rent,
- Change in median rent,
- Median home value,
- Change in median home value,
- Median household income,
- Change in median household income,
- Total number of people,
- Percentage change in number of people.

For testing purposes, I use Census-level block group data from NHGIS. Hammel and Wyly's (1996) measures that capture change over time use data from 2010 and 2018, while the measures that are simply levels are Census data from 2018.

In terms of model choice, I opt to employ Pearson's correlations to test *Hypothesis 1*. Pearson's correlation allows for researchers to examine the association between two continuous variables—in this case, the racial and ethnic composition of neighborhoods and distance to bicycle infrastructure. To be clear, Pearson's correlation tests do not allow us to examine causality, but they can tell us to what degree two continuous variables are associated with one another. When two continuous variables move perfectly together in a positive direction (when one variable increases, the other variable increases), the Pearson's correlation coefficient would be 1. When two continuous variables move perfectly together in a negative direction (when one variable increases, the other variable decreases), the Pearson's correlation coefficient would be −1. In contrast, when there is no association between the variables, the Pearson's correlation coefficient would be 0. In this case, I am interested how the racial and ethnic composition of neighborhoods is associated with distance to bicycle infrastructure. Specifically, I expect that neighborhoods (conceptualized as block groups) with higher proportions of African Americans and Hispanics will have a positive and statistically significant Pearson's correlation coefficient with distance to bikeways.

In order to test *Hypothesis 2*, I employ an ordinary least squares regression with fixed effects by city to capture the heterogeneity across these nine urban areas. The fixed effects (or a dummy variable included for each city) by city allow me to control for the different contexts across American cities. For instance, Dallas is much less dense and more suburban than New York City, so including dummy variables for each city allows for us to control for that heterogeneity. An ordinary least squares regression has one dependent variable and multiple independent variables, unlike the Pearson's correlation coefficient which is limited to simply two variables. This allows me to test how different independent variables are associated with the dependent variable, while controlling for other independent variables simultaneously. The dependent variable in this case is the relative distance to the nearest bike lane detailed earlier. The independent variables include median household income, change in median household income, percentage of persons age 25 and over with a college degree or more, change in percentage of persons age 25 and over with a college degree or more, percentage of employed people, change in percentage of employed people, median rent, change in median rent, median home value, change in median home value, median household income, change in median household income, total number of people, and percentage change in number of people. I expect that these independent variables will have negative and statistically significant associations with the dependent variable. Of course, these different facets of gentrification might have different associations with bicycle infrastructure. It should be noted that the ordinary least squares regression is not a causal test that explores whether bike lanes cause gentrification or vice versa, but rather the associations between bike lanes and the different facets of gentrification.

RESULTS

The results for testing *Hypothesis 1* can be seen in Table 8.1. While the correlation coefficients between the standardized distance measures and the standardized race and ethnicity measures are relatively low, they are statistically significant. Yet, they are significant in the direction opposite to what was hypothesized. Specifically, block groups with higher proportions of Hispanics, African Americans, and non-White groups are *more* proximate to bike lanes than block groups with higher proportions of White people. I suspect this is likely a function of the fact that much of the current bike lane infrastructure is concentrated in urban centers, which historically have had

Table 8.1 Pearson Correlation Coefficients between Distance to Bike Lanes and Standardized Measures of Race

	Distance to bike lanes
Standardized percent Black	−0.03865748***
Standardized percent Hispanic	−0.03213765***
Standardized percent non-White	−0.0182389***

Note: *p<0.1; **p<0.05; ***p<0.01
Source: This table has been generated by the author.

Table 8.2 Effects of Gentrification on Distance to Bike Lanes

	Dependent variable
	Distance to nearest bike lane
Total population (2018)	0.0001*** (0.00001)
Population pct. change	−0.197*** (0.039)
Percent employed (2018)	1.084*** (0.171)
Employed pct. change	−0.427*** (0.094)
Percent college and above (2018)	−0.760*** (0.055)
College and above pct. change	−0.011* (0.006)
Median HH income (2018)	0.00001*** (0.00000)
Median income pct. change	−0.027*** (0.007)
Median home value (2018)	−0.00000*** (0.00000)
Home value pct. change	0.010*** (0.001)
Median gross rent (2018)	0.0002*** (0.00003)
Rent pct. change	−0.005*** (0.001)
Constant	−1.134*** (0.149)
Observations	13,564
R^2	0.086
Adjusted R^2	0.085
Residual std. error	0.967 (df = 13543)
F statistic	63.911*** (df = 20; 13543)

Note: *p<0.1; **p<0.05; ***p<0.01
Source: This table has been generated by the author.

higher concentrations of people of color (Massey & Denton, 1993) or are increasingly blurring demographically (Delmelle, 2019).

The results for testing *Hypothesis 2* can be seen in Table 8.2. Because gentrification is a multidimensional concept, different facets of gentrification are related to the presence of bike lanes in different ways. First, I examine the relationship between the individual-level socioeconomic variables and bike lanes. In the case of population, block groups with higher counts of people have greater relative distance to the nearest bike lane. However, block groups that have experienced significant population growth from 2010 to 2018 have less relative distance to the nearest bike lane. A similar relationship holds true regarding percent of people employed where block groups with greater percentages of people employed have less access to bike lanes. In contrast, block groups that have experienced significant increases in the proportion of people employed have greater access to bike infrastructure. Median household income maintains a comparable relationship as well, with block groups with higher median household income having less access to bike lanes, while block groups that have experienced growth in median household income have greater access to bike lanes. Block groups that have high proportions of those with a college degree or greater and block groups that have experienced growth in the proportion of those with a college degree are more relatively proximate to bicycle infrastructure.

Second, I examine the relationship between the housing variables and the distance to the nearest bike lane. Median home value seems to decrease the distance to the nearest bike lane, suggesting that places with higher property values are associated with more accessible bicycle infrastructure. Yet, in block groups where property values have grown the most, relative distance to the nearest bike lane is, in fact, greater. Rent prices seem to have the opposite relationship, though, where block groups with higher rental prices have greater relative distance to the nearest bicycle lane. However, in communities where rental prices have increased dramatically, distance to the nearest bike lane has decreased. Ultimately, these results provide mixed support for *Hypothesis 2.*

DISCUSSION AND CONCLUSION

Ultimately, these results suggest that gentrification is, in some ways, associated with bike lanes. Because gentrification is a multidimensional concept and process, different facets had different associations with relative accessibility to bike lanes. For instance, communities where the population grew, the percent of employed people grew, the percent of people with a college education grew, and the median household income grew were relatively closer to bicycle infrastructure. This suggests that the demographic and socioeconomic changes related to gentrification are associated with having greater access to bike lanes. Conversely, when gentrification is measured in terms of current estimates—total population, percent employed, and median household income—it has a positive association with distance to bicycle infrastructure. In other words, block groups with greater populations, higher proportions of people employed, and greater median household income are going to be relatively more distant from bicycle infrastructure. It was also interesting to note that home value and gross rent had opposite relationships with distance to bike lanes. Specifically, current median home values are associated with less distance to the nearest bike lane, while current gross rent is associated with more distance. Indeed, homeownership rates—a variable not included in Hammel and Wyly's operationalization nor in the previous analyses, but nevertheless theoretically relevant—do seem to be associated with distance to bicycle infrastructure. Specifically, in a supplementary analysis, block groups with higher rates of homeownership in 2018 are more distant from bikeways than block groups with higher proportions of renters, and this association is statistically significant, according to Pearson's correlation coefficient.

These results then fit with preexisting work that demonstrates a relationship between gentrification and investments in cycling infrastructure. Specifically, scholars like Hoffman (2016) and Flanagan et al. (2017) find that gentrification and privilege are associated with bike lanes. However, this analysis extends to a greater number of neighborhoods and cities and demonstrates that different facets of gentrification are associated in different ways with cycling infrastructure. Specifically, places where there have been significant socioeconomic changes have greater access to cycling infrastructure, while neighborhoods with greater proportions of employed people and higher median household income are further from cycling infrastructure. Thus, the results herein paint a more nuanced picture of how gentrification might be associated and related with bike lanes.

These results also suggest that bike lanes in their current configuration are potentially not sited in a discriminatory way, as block groups with higher proportions of Hispanics, African Americans, and non-Whites generally were associated with more proximity to bike lanes. These results fit with work that examined minority communities' proximity to parks where disadvantaged communities tend to live closer to recreational infrastructure (Comer & Skraastad-Jurney, 2008; Jones & Coombes, 2009). Still, I would be somewhat hesitant to state that this is equitable distribution, as equity is not simply a matter of absolute distance and access. For instance, a low-income community with significant demand for bikeways may have the same access to bikeways as a high-income community with less demand. It would not be accurate to describe this as equitable, as the community with less resources *and* more demand has the same access to bikeways as the community with more resources *and* less demand. That is, equity should also be a function of need—a combination of resources and demand. Keep in mind that 0.7 percent of trips among Hispanics and 0.8 percent of trips among mixed-race groups are done by bicycle, compared to 0.6 percent among non-Hispanic Whites, suggesting that people of color have greater demand for bikeways (McKenzie, 2017). The demographics of mode choice demonstrate that some racial and ethnic minority groups are more likely to opt for bicycles than their White counterparts, suggesting that while traditionally marginalized groups are bicycling, they may not have the access to infrastructure to support their mode of transit. Moreover, accessibility in these analyses is measured as distance, while neglecting the quality of the cycling infrastructure. For instance, less politically and socioeconomically privileged communities may have inferior cycling infrastructure, compared to their wealthy, White counterparts.

Ultimately, this work presents somewhat of a paradox for activists, planners, and government officials. Clearly, gentrification and investments in cycling infrastructure are somewhat tied, though I do not speculate on the causal ordering. How then do we guarantee that marginalized groups have access to quality infrastructure, while simultaneously ensuring that these policies do not neglect or erase the preexisting residents and communities? Subsequent work should examine communities where appropriate investments have been made without harming or displacing original residents.

Finally, I propose that future research explore the temporal ordering of gentrification and investments in cycling infrastructure. In some ways, the relationship between the two likely works as a feedback process where investments in cycling infrastructure attract middle- and upper-class residents, and inflows of middle- and upper-class residents encourage public investments in cycling infrastructure. However, additional research is necessary to validate this proposition.

NOTE

1. See https://www.governing.com/gov-data/population-density-land-area-cities-map.html.

REFERENCES

Alaimo, K., Reischl, T. M., & Allen, J. O. (2010). Community gardening, neighborhood meetings, and social capital. *Journal of Community Psychology, 38*(4), 497–514.

Alaniz, M. L., Cartmill, R. S., & Parker, R. N. (1998). Immigrants and violence: The importance of neighborhood context. *Hispanic Journal of Behavioral Sciences, 20*(2), 155–174.

Blondel, B., Mispelon, C., & Ferguson, J. (2011). *Cycle More Often 2 Cool Down the Planet! Quantifying CO2 Savings of Cycling*. Brussels: European Cyclists' Federation ASBL.

Brand, C., Goodman, A., Rutter, H., Song, Y., & Ogilvie, D. (2013). Associations of individual, household and environmental characteristics with carbon dioxide emissions from motorised passenger travel. *Applied Energy, 104*, 158–169.

Bullard, R. D. (1983). Solid waste sites and the black Houston community. *Sociological Inquiry, 53*(2–3), 273–288.

Bullard, R. D., Mohai, P., Saha, R., & Wright, B. (2007). *Toxic Wastes and Race at Twenty 1987–2007: Grassroots Struggles to Dismantle Environmental Racism in the United States*. United Church of Christ Justice and Witness Ministries.

Bullard, R. D. (2018). *Dumping in dixie: Race, class, and environmental quality*. Routledge.

Checker, M. (2015). Green is the new brown: "Old school toxics" and environmental gentrification on a New York City Waterfront. In C. Isenhour, G. McDonogh, & M. Checker (Eds.) *Sustainability in the global city: Myth and practice* (pp. 157–179). Cambridge.

Cole, L. W., & Foster, S. R. (2001). *From the ground up: Environmental racism and the rise of the environmental justice movement*, 34. NYU Press.

Comer, J. C., & Skraastad-Jurney, P. D. (2008). Assessing the locational equity of community parks through the application of Geographic Information Systems. *Journal of Park and Recreation Administration, 26*(1), 122–146.

Crowder, K., & Downey, L. (2010). Interneighborhood migration, race, and environmental hazards: Modeling microlevel processes of environmental inequality. *American Journal of Sociology, 115*(4), 1110–1149.

Delbosc, A., & Currie, G. (2011). Using Lorenz curves to assess public transport equity. *Journal of Transport Geography, 19*(6), 1252–1259.

Depro, B., & Timmins, C. (2012). Residential mobility and ozone exposure. *The Political Economy of Environmental Justice, 115–136.*

Flanagan, E., Lachapelle, U., & El-Geneidy, A. (2016). Riding tandem: Does cycling infrastructure investment mirror gentrification and privilege in Portland, OR and Chicago, IL? *Research in Transportation Economics, 60*, 14–24.

Florida, R. (2005). *Cities and the creative class*. Routledge.

General Accountability Office. (1983). Siting of hazardous waste landfills and their correlation with racial and economic status of surrounding communities. United States General Accounting Office, Washington, DC.

Geoghegan, P. (2016). Blame it on the bike: does cycling contribute to a city's gentrification? *The Guardian*. https://www.theguardian.com/cities/2016/oct/05/blame-bike-cycling-contribute-city-gentrification.

Glass, R. (1964). *London: Aspects of change*. MacGibbon and Kee.

Goodman, J. D. (2010, November 22). Expansion of bike lanes in city brings backlash. *New York Times*. https://www.nytimes.com/2010/11/23/nyregion/23bicycle.html.

Greenblatt, A. (2018, June). White flight returns, this time from the suburbs. *Governing*. https://www.governing.com/topics/urban/gov-white-flight-suburbs.html.

Hamilton, James T. (1995). Testing for environmental racism: prejudice, profits, political power? *Journal of Policy Analysis and Management, 14*(1), 107–132.

Hammel, D. J., & Wyly, E. K. (1996). A model for identifying gentrified areas with census data. *Urban Geography, 17*(3), 248–268.

Herrington, C., & Dann, R. J. (2016). Is Portland's bicycle success story a celebration of gentrification?: A theoretical and statistical analysis of bicycle use and demographic change. In Golub, A., Hoffmann, M., Lugo, A., and Sandoval, G., editors, *Bicycle Justice and Urban Transformation* (pp. 32–52). Routledge.

Hoffmann, M. L. (2016). *Bike lanes are white lanes: Bicycle advocacy and urban planning*. University of Nebraska Press.

Jenkins, G. R., Yuen, H. K., Rose, E. J., Maher, A. I., Gregory, K. C., & Cotton, M. E. (2015). Disparities in quality of park play spaces between two cities with diverse income and race/ethnicity composition: A pilot study. *International Journal of Environmental Research and Public Health, 12*(7), 8009–8022.

Johansson, C., Lövenheim, B., Schantz, P., Wahlgren, L., Almström, P., Markstedt, A., & Sommar, J. N. (2017). Impacts on air pollution and health by changing commuting from car to bicycle. *Science of the Total Environment, 584*, 55–63.

Jones, A. P., Hillsdon, M., & Coombes, E. (2009). Greenspace access, use, and physical activity: understanding the effects of area deprivation. *Preventive Medicine, 49*(6), 500–505. http://doi.org/10.1016/j.ypmed.2009.10.012.

Kramer, A., & Goldstein, A. (2015). Meeting the Public's Need for Transit Options: Characteristics of Socially Equitable Transit Networks. *Institute of Transportation Engineers. ITE Journal, 85*(9), 23.

Lindelöw, D., Koglin, T., & Svensson, Å. (2016). Pedestrian planning and the challenges of instrumental rationality in transport planning: Emerging strategies in three Swedish municipalities. *Planning Theory & Practice, 17*(3), 405–420.

Litman, T. (2015). *Evaluating Public Transit Benefits and Costs*. Victoria, BC, Canada: Victoria Transport Policy Institute.

Lubitow, A., & Miller, T. R. (2013). Contesting sustainability: Bikes, race, and politics in Portlandia. *Environmental Justice, 6*(4), 121–126.

Lubitow, A., Zinschlag, B., & Rochester, N. (2016). Plans for pavement or for people? The politics of bike lanes on the 'Paseo Boricua'in Chicago, Illinois. *Urban Studies, 53*(12), 2637–2653.

Manson, S., Schroeder, J., Van Riper, D., & Ruggles, S. (2019). *IPUMS National Historical Geographic Information System: Version 14.0 [Database]*. Minneapolis, MN: IPUMS.

Massey, D. S., & Denton, N. A. (1993). *American apartheid: Segregation and the making of the underclass*. Harvard University Press.

McKenzie, B. (2017). Modes less traveled—bicycling and walking to work in the United States: 2008–2012. US Census Bureau.

Mohai, P., Pellow, D., & Roberts, J. T. (2009). Environmental justice. *Annual Review of Environment and Resources, 34*, 405–430.

Nall, C. (2018). *The road to inequality: How the federal highway program polarized America and undermined cities*. Cambridge University Press.

Neves, A., & Brand, C. (2019). Assessing the potential for carbon emissions savings from replacing short car trips with walking and cycling using a mixed GPS-travel diary approach. *Transportation Research Part A: Policy and Practice, 123*, 130–146.

Pastor, M., Sadd, J., & Hipp, J. (2001). Which came first? Toxic facilities, minority move-in, and environmental justice. *Journal of Urban Affairs, 23*(1), 1–21.

Rigolon, A., & Flohr, T. L. (2014). Access to parks for youth as an environmental justice issue: Access inequalities and possible solutions. *Buildings, 4*(2), 69–94.

Rigolon, A. (2016). A complex landscape of inequity in access to urban parks: A literature review. *Landscape and Urban Planning, 153*, 160–169.

Rigolon, A., Browning, M., & Jennings, V. (2018). Inequities in the quality of urban park systems: An environmental justice investigation of cities in the United States. *Landscape and Urban Planning, 178*, 156–169.

Ringquist, E. J. (2000). Environmental justice: Normative concerns and empirical evidence. In Vig, N. and Kraft, M, editors, *Environmental Policy: New Directions for the Twenty-First Century* (pp. 249–73). CQ Press.

Roberts, J. T., & Toffolon-Weiss, M. M. (2001). *Chronicles from the Environmental Justice Frontline*. Cambridge University Press.

Saha, R., & Mohai, P. (2005). Historical context and hazardous waste facility siting: Understanding temporal patterns in Michigan. *Social Problems, 52*(4), 618–648.

Sanchez, T. W. (2008). Poverty, policy, and public transportation. *Transportation Research Part A: Policy and Practice, 42*(5), 833–841.

Shaer, M. (2011). Not Quite Copenhagen: Is New York Too New York for Bike Lanes?. *New York Magazine*, 20.

Spotswood, F., Chatterton, T., Tapp, A., & Williams, D. (2015). Analysing cycling as a social practice: An empirical grounding for behaviour change. *Transportation Research Part F: Traffic Psychology and Behaviour, 29*, 22–33.

Taylor, B. D., Wachs, M., Luhrsen, K., Lem, L. L., Kim, E., & Mauch, M. (1995). Variations in fare payment and public subsidy by race and ethnicity: An examination of the Los Angeles Metropolitan Transportation Authority.

Taylor, B. D., & Morris, E. A. (2015). Public transportation objectives and rider demographics: are transit's priorities poor public policy?. *Transportation, 42*(2), 347–367.

United Church of Christ. Commission for Racial Justice. (1987). Toxic wastes and race in the United States: A national report on the racial and socio-economic characteristics of communities with hazardous waste sites. Public Data Access.

Vaughan, K. B., Kaczynski, A. T., Wilhelm Stanis, S. A., Besenyi, G. M., Bergstrom, R., & Heinrich, K. M. (2013). Exploring the distribution of park availability, features, and quality across Kansas City, Missouri by income and race/ethnicity: An environmental justice investigation. *Annals of Behavioral Medicine, 45*(suppl 1), S28–S38.

Vivanco, L. A. (2013). *Reconsidering the bicycle: An anthropological perspective on a new (old) thing*. Routledge.

Walljasper, J. (2012, October 18). Bicycling Means Better Business. *People for Bikes.* https://peopleforbikes.org/blog/bicycling-means-better-business/.

Willemse, L. (2013). A Flowmap–geographic information systems approach to determine community neighbourhood park proximity in Cape Town. *South African Geographical Journal, 95*(2), 149–164.

Wolverton, A. (2009). Effects of socio-economic and input-related factors on polluting plants' location decisions. *The BE Journal of Economic Analysis & Policy, 9*(1), 1935–1982.

Zavestoski, S., & Agyeman, J. (2015). Towards an understanding of complete streets. In Zavestoski, S., and Agyeman, J., editors, *Incomplete streets: Processes, practices, and possibilities* (pp. 307–315). Routledge.

Chapter 9

A Consumer Public Sphere

Considering Activist and Environmental Narratives in the Contexts of Themed and Consumer Spaces

Scott A. Lukas

In the late 2000s, Celebrity Cruises began a partnership with the environmental organization Conservation International and introduced a series of "environmentally friendly attributes" to its fleet of ships (Anon, 2009). In 2016, Royal Caribbean Cruises Ltd. (the corporation that operates Celebrity Cruises) began a similar partnership with the World Wildlife Fund (Royal Caribbean Group, n.d.). The partnership between both of these environmental organizations and the world's second largest cruise ship corporation is a surprising association given the cruise industry's notable carbon impacts on the environment.[1] The cruise ship industry has been critiqued for the practice of inappropriate waste disposal, and the ships that represent its foundations burn environmentally harmful emissions, exude toxic paints from their hulls, and contribute to noise pollution that impacts marine life (MacDonald, 2019). Given these apparent political contradictions, it is important to reflect on the circumstances that have led corporations like Royal Caribbean to develop public discourse that appears to be progressive and pro-environmentalist in nature. Royal Caribbean, like other corporations considered in this work, has begun a process of shifting its brand identity and its associated consumer lifestyle marketing to progressive and activist connotations (Holder, 2017). This shift is also noted in the many spaces that occupy its cruise ships.

In the 2000s, one of the most curious venues to have been built aboard a cruise ship was Team Earth. The former venue aboard the Celebrity Solstice version was described through its mission: "Travel meets conservation . . . where guests can raise their own eco-awareness while also learning how Celebrity's ships operate, and how their advanced systems help conserve the environment" (Anon, 2009). During its operation, the space offered visitors a series of images of nature—rainforests, endangered species, and other representations that connoted senses of calm and concern.[2] Accompanying the images were vivid titles that included "Our Forests Are Being Destroyed," "Oceans Are Warming and Are Under Siege," "Climate Change Impacts Continue to Grow," "Humanity Cannot Stand Alone," and "We Are Responsible." Beyond the visual and textual contexts, Team Earth included a set of comfortable desk and chair stations at which a computer was available for guest use. Individuals were asked to read a variety of articles and sources that focused on pressing environmental concerns, and, most interesting to this study, guests were provided a software program with which they were asked to calculate their carbon footprint. The curious tension between this environmentally focused space on a cruise ship and the fact of the cruise industry's notable environmental impacts is one that illustrates the uncanny and contradictory nature of research conducted in contemporary consumer space. Venues like Team Earth manifest the postmodern tensions, contradictions, and duplicitous alliances common in a world governed, more and more, through a consumer public sphere.

THE CONSUMER PUBLIC SPHERE

The concept of the public sphere, analyzed in great depth by the preeminent philosopher Jürgen Habermas in the 1960s, is one of the most significant foundations of contemporary theoretical work focused on the economy and the polity and their intersections in public and consumer life (Habermas, 1991). In Habermas's view, beginning in eighteenth-century Europe, one noted a shift in power and control vested in the State to forms of *Öffentlichkeit*, or public culture, in which individuals, through their participation in public venues, their engagement with news, media, and public matters of the day, began to create a critical, communicative, open, democratic, and participatory culture. The public sphere, in what some have noted is an ideal formulation (Thompson, 1995), is argued to be a space of "society engaged in critical public debate" (Habermas, 1991). While Habermas's theory of the public sphere has been subject to intense critique—including, in part, the notion that certain underrepresented groups in a given society do not have the same access to political participation as middle- and upper-class groups do—it does provide a useful foundation for this study. In particular, Habermas's emphasis on channels of communication in which political contexts are transmitted is significant to contemporary understandings of the transmission of political discourse. Contemporary revisions of Habermas's work, including concepts of "mediated publicness," provide additional contexts in which notions of the public sphere can be expanded to include the media, technological, and lifestyle dynamics of the contemporary world (Thompson, 1995).

One possible reworking of Habermas's concept of *Öffentlichkeit* is to place it within the context of the consumer world. The consumer world, which for the purposes of this work is envisioned in a post-1980s frame, is one in which individuals participate in the political and communicative channels envisioned in theories of the public sphere but in which that action takes place through various forms of technological, media, informational, and material mediation. Such mediation may include significant contexts of social media in which information is passed on, reinterpreted, and remixed through venues that include Facebook, Twitter, Instagram, YouTube, among other venues. As the Facebook-Cambridge Analytica data scandal of 2018 emphasized, the worlds of social media are particularly operative in the political and social decisions made by individuals (Kaiser, 2019). These venues of communication are also significant in the roles that they play in the worlds of themed and immersive lifestyle spaces, which are the frames of analysis of the most concern for this study.

The consumer public sphere, like the public sphere, is a political and discursive space that has major implications in terms of the nature of the polity and associated concepts, including citizenship, civics, and justice (Rawls, 1999). Citizenship, including its contexts connected to social justice and environmentalist discourse, is an especially germane focus for a consumer-focused formulation of the public sphere. As Sassen (2002) suggests, while citizenship is often envisioned as a "single concept and experienced as a unitary institution, citizenship actually describes a number of discrete but related aspects in the relation between the individual and the polity." As Sassen and others have noted, one of the articulations of citizenship that is of importance in contemporary society is between the individual and consumer institutions (Mathieu, 1999). In an era of "economic citizenship," the citizen's identity is, more and more, connected to acts of consumption: lifestyle shopping, online retail, social media participation, among other forms. The idea of citizenship being defined solely—or, at least, in large part—by economic and consumer concerns is, like the Royal Caribbean Team Earth venue, a contradictory circumstance. One challenge in the envisioning of a consumer public sphere is the circumstance in which an economic or lifestyle interest undertaken by the individual contradicts a political goal encapsulated by citizenship. As Rousseau and other have suggested in notions of the social contract, a viable democratic society is likely one in which the individual does, from time to time, relinquish her individual sovereignty as an actor for the sake of the greater good of others in the society (Rousseau, 1968).

It is in this spirit of the contradictory nature of economic citizenship in which a theory of the consumer public sphere might analyze the degree to which environmentalist, social justice, and activist political discourse can play a viable role in the consumer spaces, and the larger consumer society, in which they emerge.[3] This research is concerned with the specific ways in which such discourse is expressed, interpreted, and rechanneled within the contexts of physical or three-dimensional themed and immersive spaces. Cruise ships, theme parks, lifestyle stories, groceries, are among the many spaces that American consumers interact with in their everyday lives. These spaces are envisioned as operative ones meaning that they are impactful in terms of how they have an effect on an individual's identity, sense of self, and political consciousness. These spaces are considered through a typology of three major types of space—traditional, transitional, and experimental—which suggest tendencies of the development of environmentalist and activist discourse within space.[4] A concern of this focus is to not only analyze how such impacts take place at a micro-consumer-level but also consider how narratives of environmentalism, social justice, and those focused on concerns of the Anthropocene (the era in which the primary impacts on the environment and climate are caused by humans) have the potential to have a similar and meaningful impact on the world (Ellis, 2018; Lukas, 2018). In a time of climate change, global political and social unrest, rising authoritarianism, among other concerns, analysis of the possible ways in which the Anthropocene may be responded to in a meaningful and impactful sense is in order.

TRADITIONAL SPACES: THE MODERN POLITICS OF CONCERN

During its opening in Las Vegas in 1995, the Hard Rock Hotel and Casino offered a unique slot machine feature that has rarely been seen in other casino spaces. As described by an industry review, "Players who are bothered by any lingering moral qualms can assuage their gambling guilt instantly . . . above the slot machines is an electronic tote board, ticking away each second with the number of acres remaining in the world's rainforest" (Seal, 1995). According to the same article, an unspecified percentage of the profit from 5 of the casino's 800 slot machines would be donated to environmentalist causes. When asked about the concept behind the campaign, the Hard Rock president offered that "millions of visitors who visit Las Vegas . . . will also have the opportunity to personally do something good for the planet" (Seal, 1995). The rainforest ticker above the slot machines is a unique example of a common facet of many contemporary consumer spaces—the presence of environmental, activist, or social justice messages within contexts that traditionally do not connote these social meanings. The Hard Rock slots and their connected messaging point to a curious circumstance in the consumer public sphere in which an individual comes into contact with an environmental or political message that may appear to be completely without context or connection to the space at hand.

A number of contemporary spaces, which might be called "traditional," are noted by circumstances in which designers or operators attempt to bring an environmental or social justice narrative into the confines of the space, either as an attraction, a product campaign, a suggested practice for guests, and other variations. Monterey Bay Aquarium (Monterey, California), as an example, is a space focused on traditional goals of zoos and aquaria in terms of the presentation of various animal species and their ecosystems for public viewing. In addition to instructing guests about various conservation and research initiatives—including ocean plastics and animal species campaigns and a sea otter campaign—the aquarium operates a very popular website and program called Seafood Watch, in which guests are told about the environmental impacts of various forms of harvested seafood practice and are given recommendations about particular fish to consume (Seafood Watch, n.d.). Seafood Watch includes detailed recommendations for species from around the world and offers information about the methods used for fishing (encircling gillnets, bottom trawls, etc.). While some guests who have visited the aquarium have expressed that the information is very useful in their efforts to focus on sustainability, others have complained that the program is contradictory in terms of vegan and vegetarian principles against the eating and harm of animals, while others have said that the project is a form of moral assuagement for guests who might question the environmental value of spaces like aquaria that hold animals captive against their will.

Monterey Bay Aquarium is not unique in promoting environmentalist and social justice narratives with its guests. Theme parks, cruise ships, lifestyle stores, museums, and other spaces very commonly include messaging about charitable and environmental donations, conservation efforts, recycling programs, and many other campaigns. What distinguishes these spaces from the two other types considered in this work is the fact that such messages and efforts often appear as contradictory (such as in the case of an aquarium promoting ocean conservation while also maintaining animals in captivity), forms of moral assuagement (such as in situations in which guests might feel guilty about taking a cruise, as noted in Team Earth), and could be seen as non-impactful (such as in the Hard Rock slots and an effort that seemed to have minimal impact on the issues of the rainforests). For the consumer, the circumstance of any one of these appeals to her political, moral, or social conscience is especially complex. At any given time in a consumer space, the individual may find it challenging to locate herself in terms of historical, social, and political trajectories embedded in that space (Dickinson, 2002). Questions including "Should I visit an aquarium in the first place?" "Is it ethical to take a cruise?" "Does it matter if Disney theme parks practice recycling?" are all relevant, given the postmondern characteristics of the consumer public sphere. Given such a reality, an individual actor would be challenged to conduct the analyses suggested by C. Wright Mills (1959) in his idea of the "sociological imagination." Mills argued for the value of the individual situating his life within the historical, political, economic, and social milieu of the day, yet, as the Hard Rock slots represent metaphorically, any given moment of a consumer's life is represented by a myriad of memes, economic connections, political implications, and social connections, visible and invisible to the person.

Disney's Animal Kingdom, a zoo-based theme park, opened in Orlando, Florida in 1998. During its opening, the park was criticized for concerns related to the housing and treatment of animals in the park. People for the Ethical Treatment of Animals (PETA) and In Defense of Animals (IDA) staged protests and filed public complaints related to various species of animals held at the park (Shenot, 1995). Perhaps most troubling in terms of Disney's Animal Kingdom was evidence that thirty-one of the park's animals had died between 1997 and 1998 (Lancaster, 1998). This backdrop of ecological and conservation controversy provides a foundation for the many pro-environmental rides, attractions, statements, and policies that would follow the opening of the park. In the years following these controversies, the park has opened a number of attractions (including those discussed later) and has launched numerous campaigns that focus on what could be called positive environmental and conservation messages (Taylor, 2009). Given the park's quasi-zoo nature, some level of environmentalist discourse is a requirement, particularly given the controversies connected to other animal-focused spaces of entertainment. In 2013, the critical documentary film *Blackfish* exposed the problems associated with captive killer whales maintained in SeaWorld theme parks. In part, the film led to SeaWorld shuttering all killer whale breeding programs and performances in 2016. The Ringling Bros. and Barnum & Bailey Circus, which had been subject of multiple protests from organizations including PETA, closed its doors in 2017. It is important to note these developments as they have had impacts on many of the decisions that Disney has made in its environmentalist messaging in its parks.

Kilimanjaro Safaris, a ride-based version of a classic African outdoor safari, is an attraction that carries with it a strong conservationist message. Much of the ride is spent observing wildlife like giraffes and elephants, but it's moments in the ride in which an anti-poaching message appears—courtesy of an Audio-Animatronic elephant named Little Red—in which the ride's most interesting conservationist message to the guest is made. At the end of the ride, guests are happy to discover that Little Red has been saved from the likes of evil poachers. Interestingly enough, reported earlier versions of the ride presented a much darker reality of the social problem of animal poaching. In one version, guests were shown the sight of Little Red's mutilated carcass following the actions of poachers. According to one account, "The Imagineers [Disney ride designers] only wanted to drive home the point that killing animals is evil, but their message was too heavy-handed for a theme park and complaints were numerous at Guest Relations" (AllEars.net, n.d.). One challenge that is noted by this reaction to the ride is that traditional spaces, due to the nature of their operations and their often heavy focus on guest satisfaction, are often unable to deliver the more depressing dark and nihilistic messages that could have a great impact on guests in terms of needed transformation of their moral and political conscience (Lukas, 2015, 2016a).

Beyond this ride, Disney's Animal Kingdom includes many other attractions and even social media apps that promote an environmentalist and conservationist message. Kali River Rapids, in what appears to be a traditional water flume ride, gives guests numerous sensory opportunities to smell, hear, and view the negative effects of logging on the world's forests. Conservation Station, a child-centered area that focuses on veterinary care of animals and includes a social media app for learning about conservation efforts, is presented as a pedagogical attraction that is geared at teaching children important messages about protecting the environment and its species. Pandora, the World of Avatar, is perhaps the most curious of Disney's Animal Kingdom and its experimentation with various forms of conservationist messaging. The themeland, which opened in 2017, is based on the fictional intellectual property of James Cameron's *Avatar*. Like other Disney's Animal Kingdom themelands, it features various rides, attractions, giftshops, and restaurants, but most interesting is how this fictional world is being leveraged for its environmentalist potentials.[5] As the title of a National Geographic website post offers—"The World of Avatar at Disney's Animal Kingdom is Rooted in Real-World Conservation"—even fantasy-based or fictive science fiction worlds are argued to be spaces in which foundations of environmentalist and sustainability messages may be developed (Penning, 2018). Throughout the signature Avatar Flight of Passage motion simulation ride and in many spaces found outside the ride's queuing area, guests are invited to reflect on the plight of fictional fauna from the world of Avatar, including the mountain Banshee. Guests are told that the ecosystems of the creatures are under threat, and children are given opportunities to consult with local biologists who are purported to be experts with local Pandora species, ecosystems, and their associated threats. Critics have often argued that the pedagogical work of Disney parks illustrated in the many attractions of Disney's Animal Kingdom is fraudulent for its very fact of being presented within the material and design realms of a theme park form (Lukas, 2016c). While this aspect of the delivery of such environmentalist messaging should be critically analyzed, others could argue that such work—by its very fact of taking place within a traditional consumer space—is able to connect with individuals through contexts of entertainment and popular culture. The potential effects of such messaging in terms of the number of individuals receiving the pedagogy and in terms of the ease at which individuals may be receptive to the messaging could be much greater in terms of impacts (White, Hardisty, & Habib, 2019).

TRANSITIONAL SPACES: THE URGE FOR TRANSFORMATION

In 2015, the international coffee chain Starbucks launched an in-store and social media initiative known as #RaceTogether. Baristas who work for Starbucks are known for their signature writing of customers' names on coffee cups, and this particular campaign adopted that practice by asking baristas to write "Race Together" on the cups in an effort to "engage customers in conversation through Race Together stickers available in select stores" (BBC, 2015). Starbucks' attempt to focus on important contexts of race relations appeared to be doomed from its inception. Marketing photos of the campaign featured only white hands, while the Twitter hashtag took on unintended directions as consumers felt slighted by the campaign. One notable Tweet poked fun at the attempts of the corporation to mix coffee drinks with multicultural politics: "Mocha Money, Mocha Problems," "No Chai Left Behind," "Black Coffees Matter," and "By Any Beans Necessary" (BBC, 2015).

The Starbucks controversy reminds of another one involving Pepsi. "Live for Now," a commercial shot in 2017 and which featured celebrity Kendall Jenner, was similarly attacked for the appropriateness of recontextualizing Black Lives Matter (and the more specific incident involving an African American woman arrested during a Baton Rouge, Louisiana protest) in connection with Pepsi soft drinks. While it is true that more and more brands are engaging with activism, identity politics, and environmentalism as a way of connecting with more politically conscious consumers, critics have suggested that such appeals to these demographics amount to nothing more than pandering (Holder, 2017).

Beyond this social media campaign, the brand identity of Starbucks connotes to many a very distinctive material and interior design look. As one enters a typical Starbucks store, one is presented with both appearances of "the natural" and the authentic (images of coffee beans, nature, etc.) as well as a sense of compression in terms of the spatial distances of the contemporary world of coffee and its consumption (Dickinson, 2002). Starbucks' success has been, in no small part, due to its ability to ease many of our tensions and existential anxieties of being consumers in a world of the consumer public sphere. In a world full of fear and turmoil, the appearance of naturalness—found within Starbuck's sensory appeals of the natural wood tones of the interior design, the smell of coffee, among other examples—would seem to suggest something other than the complicated postmodern ungroundedness and arbitrariness of the world (Dickinson, 2002).[6] In Greg Dickinson's (2002) mind, the effect of such sensory design is to "cover up the difficult social, cultural and economic conditions that make the coffee available in the first place."

This duplicitous nature of Starbuck's rhetorical construction of the typical coffee consumer is a feature of the transitional space in consumer society. The etymology of "transition"—"a going across or over"—reminds of the specific challenges faced by spaces that attempt to posit a new environmental or social justice message while, simultaneously and often in contradiction, maintaining the core modes of operation as a capitalist and consumer enterprise (Online Etymology Dictionary, n.d.). Such spaces also suggest, in a more positive political sense, that certain decisions made at the corporate level and in conjunction with social and environmental concerns arising from lifestyle consumers—including operational changes to aspects of corporate chain, narratives and discourse, and practices within the spaces—may be indications of larger scale social change within society and, more specifically, its economic and consumer sectors.

One example of such a corporation that is, over time, transitioning its practices is outdoor retailer Patagonia. A quick perusal of its annual *Environmental and Social Initiatives* book, which is displayed prominently in many of its stores, including in Portland, Oregon, one is immediately confronted with contexts, examples, and rhetoric that do not match the environmental, political, economic, and social messaging of other corporations (Patagonia, 2016). As an example, the 2016 version of the initiatives book details the efforts of activists protesting on kayaks near an oil rig, lauds the desire of its founder "to use business to inspire and implement solutions to the environmental crisis," and states its corporate commitments to practices that include transparency, avoiding unnecessary harms, and providing a supportive work environment (Patagonia, 2016). Some critics have argued that the effect of this rather bold and activist rhetoric is a "greenwashing"—a ploy to maintain a loyal and committed following of upper-middle-class Patagonia consumers (Atasu, 2016; Hepburn, 2013).

A web initiative known as the Footprint Chronicles has been especially helpful as a model for future corporations that may be interested in applying a more critical analysis of their practices (Patagonia, 2020). The site includes various scorecards for all aspects of Patagonia's resource sourcing, manufacturing, labor, distribution, and consumer sales operations. Unlike many aspects of capitalist corporate practice, in which a form of invisibility of corporate operations and effects is promoted, Patagonia deliberately exposes and visualizes many aspects of its corporate operation (Rosenblum, 2012). The Footprint Chronicles (launched in 2007), perhaps envisioned as additional websites or even social media apps geared at responsible consumerism, could also be seen as a model for future corporations and consumer spaces. One of the challenges of the consumer public sphere is the relative invisibility of resource decisions, production processes, and other aspects of consumerism, and models like those of Patagonia suggest a way to open up both the corporation and consumer actors to more knowledgeable and informed decisions (cf. Lasn, 2000).

One of the enduring legacies of the activism of the 1960s was efforts made by small businesses, including independent bookstores, organic grocers, and other venues, that promoted activist political organization, labor rights for workers, and commitments to community and regional political and environmental issues (Davis, 2020). This legacy suggests that the connection of capitalist economic enterprise and activist and environmental politics may be much more of a reality (than a theoretical concept) of the consumer public sphere. Whole Foods, which was founded as a small market by John Mackey and Renee Lawson in Austin, Texas in 1978, is an example of an independent store that has become a major corporation with aspirations to maintain its more radical and political upbringings.

In 2014, the Whole Foods in Reno, Nevada created a temporary product space that was designed to resemble a traditional African thatched structure. The space included a cardboard cutout of an indigenous woman from an unspecified African culture, a sign with the words "Suppliers Investing in a Future without Poverty," a detailed map of multiple locations around the world in which Whole Foods detailed product sourcing for its stores, and numerous products, ranging from traditional Whole Foods grocery items to others like a handmade bag from Maai Mahiu, Kenya. On the surface, the temporary space appeared to display features reminiscent of a colonialist gaze as well as forms of cultural appropriation, yet, beneath these surface impressions, the contexts referenced in the displays certainly do speak of positive social justice initiatives. Of note is the map that detailed locations of product sourcing. It represents an initiative of Whole Foods' Whole Planet Foundation—an organization that like Kiva and other organizations provides micro credit opportunities for small and struggling businesses in over seventy-five countries around the world (Whole Planet Foundation, n.d.). While the political efforts of Starbucks, Patagonia, and Whole Foods will continue to be the subject of intense public scrutiny, it is important to note that transitional spaces—as ones that suggest a desire to move beyond traditional capitalist consumer foundations but illustrate a tension of being unable to fully come to terms with this shift—will continue to invite reflections on the possibilities for future engagements of space and public politics.[7]

EXPERIMENTAL SPACES: THE PROMISE OF POSTMODERNISM

The artist Bansky, who is known both for the mystery surrounding his identity and the playful political nature of his public art, opened a venue known as Dismaland in Weston-super-Mare, England in 2015. While the space was open for only one month, it received notable attention in the press (Beanland, 2015). Bansky's space featured the conceptual and installation art works of a number of political artists. Most notable among the works of art were those that mimicked the traditional rides and attractions of a theme park. Guests were invited to pilot miniature boats overflowing with refugees in "Mediterranean Boat Ride," view a SeaWorld-like whale jumping out of a toilet and through a hoop, visit a pawn shop with extremely high interest rates, and even partake in a carnival midway game of fishing for ducks in an oil-contaminated pond. A prominent yet decaying version of Cinderella Castle (invoking Disney theme park imagery) and surly and rude workers completed Banksy's vision of what many have called a dystopian theme park. While some critics have noted that Dismaland misses some of the political opportunities to critique fully consumer society and its related ills, certain exhibits and their themes, such as a large pinwheel structure intended—but unable—to power the entire park, serve as powerful metaphors of the challenges faced in a consumer world and reminders of what critical and experimental spaces of the future might imagine in terms of critique and resultant personal and social change.

Spaces like those of Dismaland may be denoted as "experimental" in that their key discursive registers identify a concern with models of cultural critique, critical play, the historical avant-gardes, postmodernism, or other sensibilities that are not represented, generally, in spaces of everyday life (Flanagan, 2009; Lukas, 2016a, 2018; Marcus & Fischer, 1999). The idea of critical play is most germane to these spaces. This form of play involves creativity and even entertainment, yet, it includes a domain of conceptual thinking about social issues and an emphasis on "an intended outcome beyond a game's entertainment or experiential value alone" (Flanagan, 2009).[8] As more and more spaces of everyday life become constituted as spaces that promote critical and goal-driven reflections, they also bring with them an emphasis on challenging the accepted (and often oppressive) social orders of the society (cf. Deleuze & Felix Guattari, 1987; Lyotard, 1984). One challenge to be noted in the aesthetic, political, and representational orders suggested in such work is the tendency for avant-garde, subcultural, and experimental lifestyle and political formations to be co-opted by mainstream media and society (Frank, 1998).

In 2015, for a period of six months, the city of Milan, Italy hosted the world's exposition. The theme of the expo was "Feeding the Planet, Energy for Life," and like other expositions of the past, the Milan version offered a number of nation pavilions, as well as thematic clusters that included Spice, Cereals and Tubers, Cocoa, among others. The Milan exposition's focus on food and positive issues of sustainability was contrasted with political concerns from protestors, including black bloc anti-austerity activists whose protests were met with tear gas and riot police efforts in the city. Prior to its opening, the exposition was marked by concerns with construction overruns, a public contract corruption scandal that resulted in numerous arrests, the use of thousands of unpaid volunteer workers, as well as public sentiment against some of the sponsors of notable pavilions, including McDonalds and Coca-Cola (Byrnes, 2015).

The world's exposition tradition is notable for its influence on popular consumer space, including the theme park industry (Lukas, 2013). It is thus not surprising that protests of the types noted were leveled against the exposition and its organizers. Though less consumerist in nature than the theme park, the world's exposition carries with it the issue of corporate sponsorship—a foundation that may be argued to be counter to concerns of sustainability (the idea of coexistence of the biosphere and humans) and environmentalism. It was interesting to note that while the Vatican had a pavilion at the event, Pope Francis, during his opening video remarks, offered a strong critique of the expo in terms of its ultimate effects: "In certain ways, the Expo itself is part of this paradox of abundance, it obeys the culture of waste and does not contribute to a model of equitable and sustainable development" (Argentieri, 2015). The pontiff's concerns suggest one challenge of consumer spaces of the future—notably, their connection to actual practices that may result in results (actual environmental or social justice enactments) and not mere consideration or rumination by the consumer.

Given many of the controversies surrounding World Expo 2015, it was surprising to see a notable dark, nihilistic, and Anthropocene-focused emphasis in a number of the pavilions. While there were numerous pavilions that focused purely on contexts of food (including in decidedly apolitical senses), and while some focused their material designs on contexts celebrating the nation being represented (some in the sense of a travel commercial for the nation), a number of the pavilions directed their emphasis on contexts of food, sustainability, and hunger in such ways that the political contexts of the representations were the major foci of the spaces. One space of this sort was the offering of Brazil. The pavilion's main structure was a giant net suspended many feet off of the ground. Guests were encouraged to walk across various portions of the structure, with instructions provided as to the easier and more challenging routes. The most interesting facet of the pavilion was the ways in which cooperation (holding hands) between guests—even strangers—made the journey across the structure much easier. The metaphor of Brazil's pavilion was clearly intended to focus concerns of food and the Anthropocene on the idea of a shared social and political journey in terms of the challenging and imminent social issues (Lukas, 2016a). Pavilion Zero, which was a non-nation-specific space near the entrance of the exposition, was one of the darkest spaces to be shown. Guests walked through a number of rooms, each of which emphasized a human effect on the environment, nonhuman species, even the psychological effects of living a consumer life. Gigantic flickering boards resembling NASDAQ stock tickers and massive piles of artistically rendered trash and garbage put the meaning of World Expo 2015 into a clear realm of that of the Anthropocene (Lukas, 2016a).

The most effective and politically engaging pavilion of the exposition was offered by the nation of Switzerland. On the exterior, the Swiss pavilion looked like many others at the expo, but what made it stand out as an experimental space was its deliberate focus on the human impact on the environment. The space offered guests a choice of taking with them one of four core food items—coffee, water, apple, and salt—each representative of a foundational aspect of Swiss food identity. Following a short discussion of the exhibit by docents, groups of visitors were given the option to take and consume any of the food items. The choice given to them was not easy. Visible from the interior and exterior of the pavilion were towers stacked with each of the items. As visitors took and consumed each of the four items, the heights of the towers decreased. Complementing the physical display in the space was a web-based counter in which each of the items was attached to a percentage. Like the physical displays, as each item was removed and consumed, the virtual tower expressed what remained (Lukas, 2016a). The Swiss pavilion was one of the most popular spaces at the exposition, and perhaps this indicates some optimism that dark thematic topics within space, including those that focus on the context of human complicity in terms of the issues of climate and sustainability, could be seen as opportunities to engage individuals within the consumer public sphere.[9]

While a majority of consumer spaces do not take the experimental route described with these spaces, the Anthropocene-focused and more critical pavilions of World Expo 2015 remind of the possibilities for spatial design in the consumer public sphere of the future. Geographer Edward Soja suggested the idea of a "thirdspace" as a concept that encapsulates both the imagination and experience of space. As he offered, "Everything comes together . . . subjectivity and objectivity, the abstract and the concrete, the real and the imagined, the knowable and the unimaginable, the repetitive and the differential, structure and agency, mind and body, consciousness and the unconscious, the disciplined and the transdisciplinary, everyday life and unending history" (Soja, 1996). Using Soja's abstract notion of space, we may imagine experimental spaces of the future which have as their key characteristic a sense of flux, fluidity, hybridity, and openness.

While some critics would argue that the expression of critical environmentalist and social justice discourse within the realm of any consumer space—whether a cruise ship, theme park, or retail store—is controversial, even unthinkable, it remains the case that more and more citizens are engaged in social media, media and technology, and consumer spaces in ways that directly relate to the frames, contexts, and foundations of the systems and structures that are responsible for social injustice and environmental harm and that are similarly connected to their unmaking, negation, and transformation.

As in the case of reframing notions of culture and heritage in the context of the Anthropocene, it may be important to consider reformulations of space given the contexts of political discourse in the consumer public sphere (Lukas, 2018). A start of that effort may include the following characteristics.

Darkness: Spaces of the future may more fully engage with nihilistic and disturbing contexts, so as to reorient the consumer to the unintended consequences and externalities of consumption practices (Lukas, 2015, 2016a).

Difficulty: Architectural approaches, such as the Reversible Destiny Foundation, have suggested new material and design practices in which spaces are created in a way in which their design provides inconvenience, not convenience, in terms of use (Lukas, 2016b). The idea behind such spaces is to use forms of spatial austerity to reorient the user of the space in existentialist and psychological senses (Lukas, 2020; Reversible Destiny, n.d.).

Spectacle: Spaces of the future, as well as their connected media and technological apparatuses, may display tendencies of spectacle (Duncombe, 2007). Specific forms of political action that may be connected to deliberate and strategic use of spectacle and public display, as in the senses of the Situationist International, could be appropriate and effect-producing in their results.

Identity: Conceptions of consumer identity—whether being studied by researchers or marketing firms—should likely be realigned to notions of postmodern and "saturated" selves (Gergen, 2000). Such new notions of identity may also include opportunities for consumers to develop reflexive and critical consciousness such that openness to the social justice and environmental concerns of the day may be manifested within the individual consumer.

A reliance on consumer choice, which seems to be the order of the day in terms of the public sphere, suggests "a starkly limited space to encourage the responsibilities and to develop the knowledge of citizenship" (Johnston, 2008). Political theorists would argue that the prerequisites of citizenship—including, most importantly, some sense of responsibility and commitment to justice for other human beings, nonhuman species, and the environment—require political, social, and economic understandings that will not likely emerge from the brand and lifestyle foci of consumer society and its associated spaces (Rawls, 1999). As Zygmunt Bauman (2009) has offered, in a world in which philosophical and ethical notions of civics and responsibility have been defined increasingly through concepts of freedom as a consumerist practice, concerns with an "Other as object of ethical responsibility and moral concern" have declined. The individuality that is often reflected in the consumer public sphere and its associated entertainment and leisure spaces is a challenging reality of a world that struggles to comes to terms with the numerous negative environmental realities of the Anthropocene. While many themed and immersive spaces do not necessarily promote material embodiments of Bauman's notion of the Other as a projection of one's conscience and altruistic sensibilities, there are a number of spaces, including some discussed in this work, that suggest an opportunity to build new communities of concern whose primary focus is something beyond the traditional materialism of consumer society.

NOTES

1. As noted in one study, "A single cruise ship can emit as much pollution as 700 trucks and as much particulate matter as a million cars" (Ellsmoor, 2019).
2. The author conducted ethnographic research in the space in 2010. All of the spaces considered in this research have been visited by the author in numerous phases of ethnographic research.
3. In this research, environmental discourse refers to focus on the environment, nonhuman species, and related areas to such an extent that the emphasis is on preservation of the environment and nonhuman species above and beyond human interests. Activist discourse refers to any emphasis on a political, economic, or cultural concern that challenges the status quo or authority structures of a society. Social justice discourse refers to any emphasis that focuses on dealing directly with pressing social problems or issues, such as racism, poverty, and many other topics.
4. It should be emphasized that this typology is provided as a theoretical foundation in which various issues of the consumer public sphere may be analyzed and that it is not intended to be applied as a practical ethnographic model.
5. It is interesting to note that visitors to Pandora are called "eco-tourists." Windtraders, a Pandora giftshop, includes recycled goods for purchase.

6. The sense of "ungroundedness" noted in theories of postmodern society suggests that individuals—as civic, consumer, and existential beings—have lost senses of authenticity that were once present in a modern world (Gergen, 2000). As well, the ungroundedness that people encounter may be attributed to conditions in which individuals experience greater "incredulity toward meta-narratives," as Jean-François Lyotard noted in his suggestions related to the decline of foundations provided by meta-narratives (politics, religion, science, etc.) (1984).

7. Another space of note that was not considered in this research is DeFiPlanet, a theme park and recreation center near Poitiers, France, which offers some very poignant and critical environmentalist narratives.

8. An additional context of such forms of play to note is the world of "serious gaming" in which traditional video games are adapted for serious and political purposes (McGonigal, 2011).

9. The contexts of the Swiss pavilion suggest some real opportunities to conduct ethnographic research related to the specific behavioral changes that are noted in consumers upon visiting such dark spaces and internalizing their messages (cf. Kozinets and Handelman, 2004).

REFERENCES

AllEars.net. (n.d.). Changes on Kilimanjaro safaris at Disney's Animal Kingdom take place this week. https://allears.net/2012/05/04/changes-on-kilimanjaro-safaris-at-disneys-animal-kingdom-take-place-this-week/

Anon. (2009, July 30). Team earth and the energy efficiencies on celebrity equinox. *Cruise Industry News*.

Argentieri, B. (2015, May 6). This is what the 30,000 protesters against the Milan Expo are complaining about. *Quartz*.

Atasu, A. (Ed.). (2016). *Environmentally responsible supply chains*. Springer. Springer

BBC. (2015, March 17). Starbucks #RaceTogether campaign mocked online. *BBC News*.

Bauman, Z. (2009). *Does ethics have a chance in a world of consumers?* Harvard University Press.

Beanland, C. (2015, September 2). Theme parks continue to draw in thrill-seekers despite the risks, so why are we so addicted? *The Independent*.

Byrnes, M. (2015, April 30). Protestors in Milan really don't want the World Expo. *City Lab*.

Davis, J. C. (2020). *From head shops to Whole Foods: The rise and fall of activist entrepreneurs*. Columbia University Press.

Deleuze, G., & Guattari, F. (1987). *A thousand plateaus: Capitalism and schizophrenia*. University of Minnesota Press.

Dickinson, G. (2002, Fall). Joe's rhetoric: Finding authenticity at Starbucks. *Rhetoric Society Quarterly, 32*(4), 5–27.

Duncombe, S. (2007). *Dream: Re-imagining progressive politics in an age of fantasy*. New Press.

Ellis, E. C. (2018). *Anthropocene: A very short introduction*. Oxford University Press.

Ellsmoor, J. (2019, April 26). Cruise ship pollution is causing serious health and environmental problems. *Forbes*.

Flanagan, M. (2009). *Critical play: Radical game design*. MIT Press.

Frank, T. (1998). *The conquest of cool: Business culture, counterculture, and the rise of hip consumerism*. University of Chicago Press.

Gergen, K. (2000). *The saturated self: Dilemmas of identity in contemporary life*. Basic Books.

Habermas, J. (1991). *The structural transformation of the public sphere: An inquiry into a category of bourgeois society*. MIT Press.

Holder, A. (2017, February 3). Sex doesn't sell any more, activism does. And don't the big brands know it. *The Guardian*.

Hepburn, S. J. (2013). In Patagonia (clothing): A complicated greenness. *Fashion Theory, 17*(5), 623–646.

Johnston, J. (2008, June). The citizen-consumer hybrid: Ideological tensions and the case of Whole Foods. *Theory and Society, 37*(3), 229–270.

Kaiser, B. (2019). *Targeted: The Cambridge Analytica whistleblower's inside story of how big data, Trump, and Facebook broke democracy and how it can happen again*. Harper.

Kozinets, R. V., & Handelman, J. M. (2004, December). Adversaries of consumption: Consumer movements, activism, and ideology. *Journal of Consumer Research, 31*(3), 691–704.

Lancaster, C. (1998, May 14). 31 animals died at Disney park. *Orlando Sentinel*.

Lasn, K. (2000). *Culture jam*. William Morrow.

Lukas, S. A. (2020, Spring). On architecture, entertainment, and discomfort. *The Right Angle Journal*, 3(6–8).

Lukas, S. A. (2018). Heritage as remaking: Locating heritage in the contemporary world. In A. M. Labrador & N. A. Silberman (Eds.) *The Oxford handbook of public heritage theory and practice* (pp. 154–167). Oxford University Press.

Lukas, S. A. (2016a). Dark theming reconsidered. In S. A. Lukas (Ed.) *A reader in themed and immersive spaces* (pp. 225–235). Carnegie Mellon University/ETC Press.

Lukas, S. A. (2016b, Fall). *Should Architecture be Entertaining? The Architecture of Entertainment II*. OAA Perspectives, Ontario Association of Architects, 22–23.

Lukas, S. A. (2016c). Judgments passed: The place of the themed space in the contemporary world of remaking. In S. A. Lukas (Ed.) *A reader in themed and immersive spaces* (pp. 257–268). Carnegie Mellon University/ETC Press.

Lukas, S. A. (2015, Quarter 4). Controversial topics: Pushing the limits in themed and immersive spaces (Masterclass). *Attractions Management, 20*, 50–54.

Lukas, S. A. (2013). How the theme park got its power: The world's fair as cultural form. In C. Pearce, et al. (Eds.) *Meet me at the fair: A World's Fair reader* (pp. 395–407). Carnegie Mellon University/ETC Press.

Lyotard, J. F. (1984). *The postmodern condition: A report on knowledge*. University of Minnesota Press.

MacDonald, J. (2019, July 1). The high environmental costs of cruise ships. *JSTOR Daily*. https://daily.jstor.org/the-high-environmental-costs-of-cruise-ships/

Marcus, G. E., & Fischer, M. M. J. (1999). *Anthropology as cultural critique: An experimental moment in the human sciences*. University of Chicago Press.

Mathieu, P. (1999, Autumn). Economic citizenship and the rhetoric of gourmet coffee. *Rhetoric Review 18*(1), 112–127.

McGonigal, J. (2011). *Reality is broken: Why games make us better and how they can change the world*. Penguin.

Mills, C. W. (1959). *The Sociological Imagination*. Oxford University Press.

Online Etymology Dictionary. (n.d.). Transition. https://www.etymonline.com/search?q=transition.

Patagonia. (2020). Environmental and social footprint. https://www.patagonia.com/our-footprint/

Patagonia. (2016). Environmental and social initiatives. https://issuu.com/thecleanestline/docs/patagonia-enviro-initiatives-2016.

Penning, M. (2018, March 2). Connect to protect: How Pandora—the world of avatar at Disney's animal kingdom is rooted in real-world conservation. *National Geographic*. https://blog.nationalgeographic.org/2018/03/02/connect-to-protect-how-pandora-the-world-of-avatar-at-disneys-animal-kingdom-is-rooted-in-real-world-conservation/

Rawls, J. (1999). *A theory of justice*. Belknap.

Reversible Destiny. (n.d.). http://www.reversibledestiny.org.

Rosenblum, J. (2012, December 6). How Patagonia makes more money by trying to make less. *Fast Company*.

Rousseau, J. J. (1968). *The social contract*. Penguin.

Royal Caribbean Group. (n.d.). Community. https://www.rclcorporate.com/community/.

Sassen, S. (2002). The repositioning of citizenship: Emergent subjects and spaces for politics. *Berkeley Journal of Sociology, 46*, 41–66.

Seafood Watch. (n.d.). https://www.seafoodwatch.org.

Seal, K. (1995, March 20). Hard rock slots have a conscience. *Hotel & Motel Management, 3*, 48.

Shenot, C. (1995, December 10). The captivity question Disney's proposed park makes an attractive target for animal-rights groups. *Orlando Sentinel*, 9.

Soja, E. (1996). *Thirdspace: Journeys to Los Angeles and other real-and-imagined places*. Basil Blackwell.

Taylor, B. (2009). *Dark green religion: Nature spirituality and the planetary future*. University of California Press.

Thompson, J. B. (1995). *The media and the modernity: A social theory of the media*. Polity.

White, K., Hardisty, D. J., & Habib, R. (2019, July-August). The elusive green consumer. *Harvard Business Review*. 124–133

Whole Planet Foundation. (n.d.). https://wholeplanetfoundation.org.

Chapter 10

What Role Can Water Markets Play in Adapting to Climate Change?

Evidence from Two River Basins in the Western United States

Elizabeth Koebele, Loretta Singletary, Shelby Hockaday, and Kerri Jean Ormerod

INTRODUCTION

Across the globe, climate change is reshaping environmental governance. Policy-makers and environmental managers can no longer rely on past trends to inform their decisions; instead, they must plan for an uncertain future characterized by increased climate variability and temperature extremes that lie outside of the observed record (Woodhouse et al., 2016). Even if countries across the globe begin taking coordinated action to mitigate climate change right now, many of its impacts will continue to be felt for decades to come (Adler, 2012; U.S. Global Change Research Program, 2018). Thus, to moderate these impacts, societies must *adapt* or permanently adjust their "business-as-usual" processes in response to projected or current climate-driven changes (Intergovernmental Panel on Climate Change, 2015).

The semi-arid western United States is expected to experience some of the most severe impacts of climate change. Because this region, like many others throughout the Northern Hemisphere, is dependent on winter snowpack to "store" water for warmer and drier summer months, anticipated shifts in the region's precipitation patterns are expected to have significant and cascading socio-environmental impacts (Mankin et al., 2015). The best available science projects more variable precipitation and rising temperatures across much of the region (Intergovernmental Panel on Climate Change, 2015; U.S. Global Change Research Program, 2018). This will cause more winter precipitation to fall as rain rather than snow, leading to a smaller snowpack that melts earlier in the spring (Regonda et al., 2005). Such changes will alter the quantity and timing of spring runoff, in turn creating capacity and conveyance challenges for water infrastructure that was established decades, if not a century, ago. Higher temperatures are also expected to intensify human and environmental demand for water (Lall et al., 2018), further stressing water management systems that already precisely monitor and distribute water for multiple uses under historical conditions of scarcity.

Many of the deleterious impacts of climate change on water supplies in the western United States are already occurring (Barnett et al., 2008). Scientists have detected reductions in Colorado River flows that are due to a smaller winter snowpack and higher evapotranspiration rates (Milly & Dunne, 2020; Udall & Overpeck, 2017). These kinds of pervasive drought conditions, which have been occurring since the early 2000s, have already exacerbated water shortages across economic sectors in the region and caused significant environmental impacts. For instance, during a recent drought, the nation's agriculture sector lost approximately $35 billion in output, with the state of California suffering the largest loss at roughly $1.5 billion in 2015 (Ghosh, 2019). It is evident that this region must prepare to adapt to current and projected changes to livelihoods, the natural environment, human health, and overall quality of life.

When considering adaptation approaches in the region's water resources sector, it is crucial to understand that most available water is already fully allocated to water users through "water rights," which are governed under a legal doctrine called prior appropriation (Tarlock, 2002). Under this doctrine, water rights are viewed as property rights that may be bought and sold separately from the land on which they were first utilized. While this arrangement theoretically facilitates the easy exchange of water rights, such transactions have been relatively limited for a number of reasons, including a lack of information about willing buyers or sellers, high administrative costs to approve transfers, and the preference of water rights holders to retain ownership of their water rights (Ghosh, 2019; Squillace, 2013). Prior appropriation also dictates that in times of water shortage, water rights are curtailed based on seniority—meaning that "junior" water rights established later in time are "cut off" to ensure that older "senior" water rights are fulfilled, without regard to the purpose for which the water is used. As such, the prior appropriation doctrine has been criticized for locking in "outdated" water uses, such as water-intensive irrigated agriculture that came to the region early in its development (Tarlock, 2002).

Given this context, developing mechanisms to more flexibly reallocate water is necessary, particularly in light of changing social values, shifting regional economies and demographics, and water supply volatility. Of the potential adaptation strategies readily available, *water markets* have gained increasing attention from water resources scholars and managers alike. Water markets facilitate voluntary exchanges of quantifiable amounts of water among willing buyers and willing sellers. These exchanges include temporary leases of water, as well as permanent acquisitions, depending on the goal and design of the market (Culp et al., 2014). In regions such as the western United States where established water rights exist, water markets can help move water to where it is economically valued most under changing supply-and-demand conditions. Through price signals, markets can encourage water conservation and motivate new investments in water storage and conveyance infrastructure (Libecap, 2018), which can help the region adapt to climate change. Thus, water markets are a promising climate adaptation measure in the western United States and other semi-arid, snow-fed regions with clearly defined water rights because they help facilitate water reallocation under uncertain supply conditions (S. E. Anderson et al., 2019; Colby & Isaaks, 2018; Garrick et al., 2018; Ghosh, 2019; Jones & Colby, 2010).

Importantly, however, water markets are not a panacea for water challenges in the western United States. Aridity is a defining feature of the region, which suggests that its residents have always faced some degree of water scarcity and will likely continue to do so. Thus, while water reallocation may help to solve supply-and-demand issues on a temporary basis, there is always going to be hard limit to the total amount of water available. Moreover, numerous critiques of water markets exist, such as their potential to create negative externalities like distributional inequity, and undermine social, economic, and ecological resiliency (Adler, 2012; Boelens & Zwarteveen, 2005). Water transfers can also be difficult to monitor and account for, given the fluidity of public information available, among other administrative challenges (Colby & Isaaks, 2018; Holley & Sinclair, 2018). These critiques suggest that effective water markets are those that can, under variable supply conditions, maximize information exchange and minimize coordination challenges while attending to changing societal values (Garrick et al., 2018).

Given the growing attention to water markets in the western United States and other areas around this globe, this chapter investigates the following question: *How can water markets be used to adapt to hydroclimatic changes in semi-arid, snow-fed river basins in the western United States?* To answer this question, the chapter proceeds in two main parts. Part 1 describes the history of water development in the western United States and considers how water markets fit into the region's existing structure for allocating scare water resources. Next, it highlights specific ways in which water markets may help water users adapt to changes in climate and water supply in this context, as well the limitations water markets may pose. Part 2 then presents two case studies of river basins in the western United States with active water markets.

In the South Platte River Basin in Colorado, a well-established water market facilitates the purchasing and leasing of water diverted from the Colorado River to the Eastern Slope of the Rocky Mountains for a variety of uses. In the Walker River Basin, located in west-central Nevada, market-based programs facilitate both temporary and permanent transfers of agricultural water rights to enhance environmental instream flows. In describing these cases, we aim to depict different approaches to designing and implementing water markets. We conclude by speculating on the challenges and opportunities these cases present to draw lessons about the use of markets as a climate adaptation mechanism broadly.

PART 1: WESTERN WATER MANAGEMENT AND WATER MARKETS

Water Development in the Western United States

During the nineteenth century, an unprecedented period of migration to the semi-arid public lands west of the 100th meridian[1] occurred. The U.S. federal government offered various federal initiatives to lure settlers westward and encourage development of the region through small family farms (Tarlock, 2002). Perhaps the most well-known of these initiatives was the Homestead Act of 1862, which encouraged settlement of the region by providing individuals with homesteads, or 160-acre parcels of public land, at little or no personal cost. Similarly, the Desert Land Act of 1877 aimed to develop large sections of arid public lands by allowing settlers to file claims to 640-acre land tracts, contingent upon proof that claimed lands had been irrigated (Ganoe, 1937; Landstrom, 1954). Early explorers such as William Gilpin endorsed these federal programs, claiming that the lands west of the 100th meridian were "no desert, nor even a semi-desert, but a pastoral Canaan" (Stegner, 1954, p. 3). According to Gilpin, settlers would thrive thanks to effortless agriculture and plentiful resources, attain heightened health and well-being in the region's temperate climate, and discover independence and freedom in a vast and open land.

Despite these attractive appeals, few settlers succeeded in farming the semi-arid lands of the western United States. In addition to encountering harsh temperatures, they discovered a water supply that paled in comparison to that of the lush eastern United States. Instead of receiving ample rain throughout the year to sustain agriculture, the western United States depends primarily on the accumulation of snowpack at high altitudes during the winter months, which then melts and runs off through rivers at lower elevations during the spring and summer (Diffenbaugh et al., 2013). Together, this unique geography and hydroclimate create a number of constraints that had to be accounted for in developing and governing western water resources.

First, it was judicially determined that the water allocation scheme used in the eastern United States, called riparianism, would not meet the needs of the comparatively water-scarce region (Leonard & Libecap, 2019). Under riparianism, only owners to first establish claims to land parcels adjacent to streams could divert water for their private and "reasonable use" (Lueck, 1995). This system, if applied in the semi-arid West, would leave most early settlers without water to sustain their newfound livelihoods. The doctrine of prior appropriation was consequently put into place in most western states and persists to this day (Tarlock, 2002). Initially, a seemingly straightforward means of allocating water to early settlers, the doctrine of prior appropriation has evolved into a complex administrative system to allocate water rights and adjudicate competing claims to scarce water resources (Garrick et al., 2018).

The prior appropriation doctrine is founded on the principle of priority or "first in time, first in right." This principle establishes a hierarchical order for claims from a given water source based on the date of the original claim. Each water right defines the water source, the amount of water that can be diverted annually, and the specific purpose and location where the water can be used. During a water-short year (i.e., times of drought), the total amount of water rights often exceeds the total water supply available. In this case, water deliveries are curtailed based on the total amount of water available and priority date, with the most junior rights potentially receiving no water (Lee et al., 2020). Therefore, junior users generally face a less secure water supply and receive less water on average than senior users (Burness & Quirk, 1979; Libecap, 2011).

Under the doctrine of prior appropriation, water must also be applied to a beneficial use. Beneficial uses were historically defined as uses that diverted water from rivers and applied it elsewhere in an economically productive manner, such as irrigated agriculture, other industry, or domestic supplies. Moreover, an anti-speculation tenet of prior appropriation, colloquially referred to as "use it or lose it," was designed to promote the full use of available water for beneficial uses by asserting that a water rights holder would lose all or part of their water right if it was not used regularly. While theoretically developed to prevent hoarding and wasteful use, this tenet had the unintended consequence of encouraging water rights holders to divert their full allocation, even if unnecessary, in order to secure their right to future use. In the past several decades, various states have revised their beneficial use statutes to include nonconsumptive uses that leave water in the stream (Schilling, 2018). These "instream flows" can enhance riparian health and water quality, restore or sustain riparian environments and wildlife habitat, and also provide for recreational uses.

In addition to requiring a new, priority-based scheme for allocating scarce water resources, the hydroclimate and geography of the western United States also prompted the construction of significant water infrastructure to store and convey water supplies over long distances for the benefit of settlers, who were primarily ranchers and farmers (Leonard & Libecap, 2015). Because on the seasonally dependent nature of western water supplies—as well as the concept that water rights function like private property rights that can be bought and sold—individual water rights holders were incentivized to collectively invest in water infrastructure in order to make full use of their water rights. These investments consequently increased the value of their lands and water rights under uncertain hydrologic conditions (Leonard & Libecap, 2019). Later, the U.S. government-funded large-scale water infrastructure development through the landmark National Reclamation Act of 1902. This legislation mandated the federal government to fund irrigation projects to aid settlers in reclaiming the deserts of western United States, which was seen as a direct means to reinforce individual enterprise rooted in agrarian-based economies and values (Akhter & Ormerod, 2015). It also led to the establishment of the U.S. Bureau of Reclamation (formerly the Reclamation Service), which coordinates infrastructure development and the transfer of water across river basins today.

A Role for Water Markets within Prior Appropriation?

The water allocation institutions and infrastructure described above were developed when the western United States was relatively sparsely populated and settlers had relatively few demands for water. Now, their continued viability is being challenged by current and projected changes in water supply due to climate change, as well as demographic shifts and evolving social values. Indeed, irrigated agriculture remains the primary economic driver in many parts of the western United States, generating billions of dollars in annual income (Culp et al., 2014; Tarlock, 2018) and using about 80 percent of all water consumed in western states (Colby & Isaaks, 2018), which is priced comparatively lower than nonagricultural water (Brewer et al., 2008; Glennon, 2009; Libecap, 2018). However, water scarcity associated with droughts, coupled with increased competition for water resources, has wrought an era of water *re*allocation. For instance, as alluded to above, many states have created mechanisms to reallocate water from agricultural to environmental purposes, despite its omission from historical beneficial use statutes (Schilling, 2018; Smith, 2019). Similarly, growing water demand in urban areas has resulted in pressures to reallocate larger quantities of water from irrigated agriculture to municipal uses (Schwabe et al., 2020).

Numerous scholars suggest that water reallocation can be done most practically and efficiently through water markets, which facilitate voluntary transactions between willing buyers and sellers. These transactions include the permanent transfers of water rights—as has always been allowed under prior appropriation—as well as short- or long-term leases of water and water storage (Brewer et al., 2008; Colby & Isaaks, 2018; Culp et al., 2014; Leonard & Libecap, 2019; Libecap, 2018; Schwabe et al., 2020). While water markets may operate differently concerning methods for information exchange, transactions, and water-pricing, all share a common goal of reallocating water to where it is most valued most (Brennan, 2017; Clifford et al., 2004).

Following basic economic principles, water markets have the potential to improve water-use efficiency and facilitate tradeoffs across various sectors via price signals (S. E. Anderson et al., 2019). By connecting willing buyers and sellers, markets can help to reduce the costs involved with buying, selling, or trading water (i.e., transaction costs). Markets can also enhance water supply reliability for junior users that are regularly curtailed (O'Donnell & Colby, 2010), incentivize coordinated demand management, and contribute substantively to reallocation by generating information about alternative and highest-valued water uses. In these ways, water markets create flexibility for water rights holders to respond to changes in supply and demand.

Some argue that the widespread use of water markets is unnecessary because the doctrine of prior appropriation is already sufficiently flexible in that it allows for permanent water right transfers (Schutz, 2012; Lee et al., 2020). However, permanent transfers have been limited for a variety of reasons, including that the approval process can be time-consuming and costly because it seeks to fully protect existing water rights holders by requiring sellers to prove that no incidental harm will occur to users as a result of the transfer. Permanent transfers also tend to be less desirable to those who want to maintain existing water rights and may even promote waste of water that could be leased to others who need it due to the "use it or lose it" tenet of the prior appropriation doctrine. For these reasons, administering the prior appropriation doctrine as it currently exists may "raise the costs today of reallocating water to higher-valued uses and of flexibly responding to hydrological uncertainty due to climate change" (Culp et al., 2014).

Current and Future Uses of Water Markets

Both formal and informal water markets already exist in most western states (Brennan, 2017; Libecap, 2018). These markets vary in complexity and include options such as water rights sales, renting or leasing, and managed aquifer storage and recovery (i.e., water banking) (Ghosh, 2019). They also vary in size from trades between users in a single irrigation district to trades across state lines (O'Donnell & Colby, 2010). There is evidence that these markets have created a variety of net social gains. For example, over the past several decades, water markets have helped secure water supplies for cities in the urbanizing western United States (S. E. Anderson et al., 2019; Ojha et al., 2018). They have also been successful in incentivizing agricultural water conservation and reallocating water to address drought-induced water shortages (Howitt et al., 2012). Severe and pervasive drought impacts in recent years have also brought about structural market innovation intended to increase trading activity in Texas, Colorado, California, and Arizona, which accounts for 60 percent of all water trades in the western United States (Ghosh, 2019). Additionally, rising market prices for water may signal a growing recognition of the potential opportunities presented by water markets under changing supply-and-demand circumstances. For example, in Nevada's Truckee River Basin, between 2002 and 2009, the median price for an acre-foot of water transferred from agriculture to municipal use was twelve times the price for the same amount traded within the agricultural sector (Libecap, 2011).

Despite the many opportunities associated with the markets described earlier, common criticisms of water markets are important to consider when evaluating their role as a potential climate adaptation strategy. For instance, concerns about expanding water markets involve existing storage and delivery infrastructure limitations, transaction costs involved in proving no injury to others as a result of transfers, and the need for a more transparent information flow to inform market behavior (Colby & Isaaks, 2018; Libecap, 2018). These concerns resonate with broader critiques of the ability of markets to promote equity among water users. For example, historically marginalized, disempowered and/or low-income water users may be unable to afford water to meet their basic needs. Furthermore, through traditional cost-benefit analysis, it can be difficult to quantify the nonmarket aspects water use, such as the ways water supports the environment or improves aesthetics (Boelens & Zwarteveen, 2005; Wattage et al., 2000). Fortunately, these issues can be addressed to some degree by designing water markets that create and enforce operational rules of exchanges (Garrick et al., 2018), including, for example, rules about reallocating water rights specifically for environmental or municipal uses (Adler, 2012; Grafton et al., 2011; Libecap, 2018; Murphy et al., 2009).

Additionally, while there are many examples of water markets in western states, large-scale, facilitated water trading remains uncommon in the region, with less than 2–4 percent of annual water consumption traded (Libecap, 2018). This has been the case despite irrigated agriculture continuing to represent about 60–80 percent of all water use across the region (Brewer et al., 2008) and increasing water scarcity for all users. Minimal trading is due to the relative novelty of markets and the limited experience of potential users, perceived sociopolitical threats to the status quo, negative externalities as a result of transfers, a lack of information, and prohibitive transaction costs—all factors that may drive up water market prices when limited opportunities are available (Clifford et al., 2004; Dilling et al., 2019; Jones & Colby, 2010; Smith, 2019; Squillace, 2013). These challenges to market efficiency must be addressed before water markets can be considered a viable, large-scale climate adaptation strategy in the western United States.

PART 2: CASE STUDIES OF SELECT WATER MARKETS IN THE WESTERN UNITED STATES

To further explore the viability of water markets as a climate adaptation strategy in the western United States and similar regions, we present case studies of two snow-fed, semi-arid river basins and their current water market programs. These cases allow for the exploration of multiple water market designs and speculation about the potential opportunities and challenges associated with them for adapting to changing hydroclimatic conditions.

Water users in both case study basins—the South Platte River Basin in northeastern Colorado, and the Walker River Basin in west-central Nevada—face spatial and temporal water allocation challenges. In an attempt to address these historical challenges, which are likely to be exacerbated under climate change, entities in both basins have instituted market-based mechanisms to reallocate water. In the South Platte River Basin, the Colorado-Big Thompson (C-BT) Project has diverted water from the Colorado River, through the Rocky Mountains, to supplement water supply on Colorado's Front Range since the mid-1900s; this water is allocated to users via a number of market-based purchasing, leasing, and storage mechanisms. In the Walker River Basin, a major federal environmental restoration program has created market-based mechanisms through which permanent and temporary water transfers occur for the purposes of environmental restoration. Both basins and their markets are discussed in detail below. Table 10.1 summarizes the major characteristics of each case study's geographical, hydrologic, and institutional contexts, as well as the water market mechanisms that are currently in place.

Case Study 1: The South Platte River Basin

The South Platte River Basin is located on the Eastern Slope of the Rocky Mountains, primarily within the state of Colorado. Rangeland dominates the

Table 10.1 Comparison of the South Platte and Walker River Basin Case Studies

	South Platte River Basin	*Walker River Basin*
Drainage area	~24,300 square miles	~3,950 square miles
Geographic context	Local precipitation falls as snow on the Eastern Slope of the Rocky Mountains and flows across Colorado's northeastern plains. The Colorado-Big Thompson (C-BT) Project provides for supplemental water diverted from the Colorado River Basin on the Western Slope of Rocky Mountains to the South Platte River Basin.	Precipitation falls as snow in the Sierra Nevada range in east-central California and the Sweetwater range along the California-Nevada border. Surface runoff and groundwater recharge flow northeast into west-central Nevada, creating the east and west forks of the Walker River. The river system flows through several major agricultural valleys, including Walker River Paiute Reservation, before terminating in Walker Lake.
Primary water uses	The majority of water within South Platter River Basin, including C-BT water, is used for irrigated agriculture, with second highest use being municipal for the Denver-metro area; some water is allocated for endangered species habitat via federal water rights.	The dominant use of Walker River water is irrigated agriculture. A distant second use is environmental instream flows to Walker Lake to recover a threatened trout fishery and migratory waterfowl habitat.
Water management and institutional norms	The prior appropriation doctrine governs the allocation of surface and tributary groundwater (i.e., groundwater that is hydrologically connected to surface water). Water is managed by water commissioners overseen by the Colorado Division of Water Resources.	The prior appropriation doctrine governs the allocation of surface and groundwater. The Walker River's surface flows are managed by the basin's federal water master; primary and supplemental groundwater is managed at the state level by the Division of Water Resources.

(Continued)

	South Platte River Basin	*Walker River Basin*
Description of Market(s)	The Colorado-Big Thompson (C-BT) Project provides supplementary water to water rights holders in the South Platte River Basin. Water rights holders (agriculture, municipal, industrial) within the boundaries of the Northern Colorado Water Conservancy District, which oversees the Project, may purchase "allotments," or shares, of C-BT Project water, which is delivered on an annual basis. Allottees may then lease portions of their annual water allotments to other rights holders, store portions of unused water in Project reservoirs, and also buy and sell storage that is available, all on a year-to-year basis.	(1) The Stored Water Program, which is administered by the Walker River Irrigation District (WRID) and funded by the National Fish and Wildlife Foundation (NFWF) aims to demonstrate how temporary transfers of surplus water stored in system reservoirs can enhance environmental instream flows to Walker Lake. (2) The Walker Basin Conservancy, an out-growth of NFWF, was established in 2015 as a 501(c)(3) nonprofit to facilitate the permanent acquisition and transfer of water rights for environmental restoration of Walker Lake. (3) Irrigators with surface water rights can buy and sell rights to storage water on a year-to-year basis through WRID.

Source: This table has been created by the authors.

landscape, except for the Denver metropolitan area, one of the fastest growing urban regions in the western United States. The Basin's major water source is the South Platte River, which flows northeast from the Continental Divide across the Colorado plains before reaching its confluence with the North Platte River in Nebraska (Guy et al., 2011). Downstream from the Denver metropolitan area, irrigated crop and livestock production comprise 37 percent and 41 percent of land use, respectively (South Platte Coalition for Urban River Evaluation, 2015). While municipal and industrial water use has increased along with steady population growth in the metropolitan area (6 percent of total demand), agricultural water use continues to dominate at 86 percent of total water demand (South Platte Basin Roundtable, 2019).

The South Platte River Basin's average annual streamflow is approximately 1.4 million acre-feet[2] (Water Education Colorado, 2019). In any given year, three potential sources of water supplies exist: (1) South Platte River surface flows that originate on the Eastern Slope of the Rocky Mountains; (2) interbasin transfers, including the C-BT transmountain diversion project, that provide supplemental flows; and (3) hydrologically connected groundwater resources. As is the case in much of the western United States, the South Platte River's waters are over-allocated, leading to concerns about future water supply, especially as climate change portends radical changes in the timing, type, and amount of total annual precipitation.

According to 2050 water-use projections made by Colorado's primary water governance agency, agriculture will continue to make up 82 percent of consumptive use (Colorado Water Conservation Board, 2011). Municipal and industrial use is projected to increase demand to levels that cannot be fulfilled by the South Platte River based on its current perennial yields. A significant number of water rights in the basin have already been permanently transferred from agricultural uses to municipal and industrial uses that supply water to homes and businesses. As of 2010, municipal and industrial users comprised about 65 percent of water rights ownership in the basin. Climate change is concurrently impacting water supply and quality in the basin, with reduced annual precipitation and increased groundwater pumping for consumptive use. These changes affect human water users and wildlife species alike (National Research Council, 2005).

The Colorado-Big Thompson Project

In 1937, the U.S. Bureau of Reclamation began building the Colorado-Big Thompson (C-BT) Project to supplement a chronic shortage of South Platte River flows needed to irrigate about 640,000 acres of agricultural lands (Howe, 2011). The C-BT Project, managed by the Northern Colorado Water Conservancy District (from here, "Northern Water"), transfers water from the headwaters of the Colorado River Basin on the Western Slope of the Rocky Mountains to the Big Thompson River on the Eastern Slope through a 13.1-mile long tunnel (Northern Colorado Water Conservancy District, 2020b). Transferred water is stored in twelve surface reservoirs and distributed to South Platte Basin water users through both Northern Water-owned and privately-owned canals and ditches. While C-BT Project water was initially intended to supplement irrigated agriculture within the basin, today it also provides water for municipal and industrial uses and generates hydropower via six hydroelectric power plants. Reaching about 960,000 people, the C-BT Project collects and delivers over 200,000 acre-feet of water every year (Northern Colorado Water Conservancy District, 2020b).

Existing water rights holders on the Eastern Slope who want to obtain supplemental C-BT Project water must first purchase an "allotment" from a willing seller, which essentially represents a percentage-based share of the total water delivered via the C-BT Project in a given year. The total amount of water available to those holding allotments (i.e., allottees) is called the water delivery quota, which is calculated each November and April by the Northern Water Board of Directors (Northern Colorado Water Conservancy District, 2020c). The quota is determined based on a variety of factors, including measurements of winter snowpack and assessments of existing water storage in project reservoirs.

Allottees may participate in a number of market-based programs that allow them to permanently sell their allotment, store a limited portion of their allotment in Project reservoirs for use in the next year, or even lease additional "carryover [storage] capacity" from other users on a year-to-year basis (Northern Colorado Water Conservancy District, 2020a). Allottees may also lease their unused water to other water users within Northern Water boundaries (called "rental water") on an annual basis for a fairly low transaction cost (Carey & Sunding, 2001). Involved parties voluntarily negotiate and settle on price, and all leases must be reviewed and approved by Northern Water and meet the stipulation for beneficial use. Finally, in years when Project reservoir storage capacity exceeds a certain amount, Northern Water implements an auction system through which water users can bid on additional C-BT Project water.

In recent years, municipal and industrial users have been purchasing excess C-BT Project allotments previously held by agricultural users to "stock up" for their immediate and expected future needs, leading to a situation in which some of the agricultural users who previously sold their allotments to municipal and industrial users now lease that same water back for irrigation on an annual basis (Carey & Sunding, 2001). This leasing market mechanism and the low cost of C-BT Project water in general has allowed for agriculture to remain the majority user of all water transferred from the Colorado Basin into the South Platte Basin (Maas et al., 2017), even as municipal and industrial users now make up the majority of C-BT Project allottees. That said, C-BT water prices have increased significantly in recent years due to a number of factors such as drought and decreased agricultural market prices, moving more water out of reach for farmers. Minimum bids for one acre-foot of water in 2010 successfully sold at $22, but in 2018 increased sixfold to $132 (Lounsberry, 2019). Despite these changes, diverse water users in the basin support water trading as part of an integrated water resource management approach that meets their increasing supply-and-demand gap while combating the trend of "buying and drying" agricultural lands (South Platte Basin Roundtable/Metro Basin Roundtable, 2015).

Case Study 2: The Walker River Basin

The Walker River Basin encompasses approximately 3,950 miles (Lopes & Allander, 2009), about 25 percent of which is located in California where the east and west forks of the Walker River originate as winter snowpack at high elevations in the Eastern Sierra Nevada range and Sweetwater Range on the California/Nevada border (California Department of Water Resources, 1992; Horton, 1996). The Walker River's average annual streamflow is approximately 387,000 acre-feet per year (Lopes & Allander, 2009). The river and its tributaries flow eastward through five distinct irrigated agricultural regions, including Bridgeport and Antelope valleys in California, and Smith Valley, Mason Valley (where the east and west forks merge), and the Schurz area in Nevada before reaching Walker Lake, a terminal deep-water lake (Tracy, 2004) and one of the few natural perennial lakes in the Great Basin (Lopes & Allander, 2009). The Walker River Paiute Reservation, located at the northern edge of Walker Lake, comprises about 330,000 acres (Begay, 2018), but does not include Walker Lake.

The Walker River has been diverted since the mid-nineteenth century, first to support mining and then to support primarily cattle and hay production (Lopes & Smith, 2007). Water diverted for irrigation continues to support the basin's agricultural economy (U.S. Bureau of Reclamation, 2010). While Walker River Paiute Reservation lands are also used mainly for grazing and ranching, most individual agricultural operations are small, as the reservation was subdivided into 20-acre allotments as a result of the Dawes Act (1887), which forced tribal communities to allocate their reservation lands to individual members specifically for agricultural enterprises (Colby et al., 2005; Cosens & Royster, 2012). Thus, most of the reservation's riparian areas have been converted to irrigated farmland, and alfalfa hay remains the major crop (Tiller, 2006).

Because of the interstate nature of the Walker River Basin, its surface waters were allocated through a 1936 interstate decree, known as the Walker River Decree, in the Federal District Court for the District of Nevada.[3] The prior appropriation doctrine is the basis for water allocation under the Walker River Decree. Decree rights are allocated by the river's federal water master, who is appointed by the U.S. Board of Water Commissioners. In 1940, the decree was modified to establish a "Federal Indian reserved water right" with an 1859 priority date[4] for the Walker River Paiute Tribe to divert 9,425 acre-feet for a period of 180 days in order to irrigate approximately 2,100 acres of lands on their reservation (Nevada Division of Water Planning, 1999), though litigation over the quantification of this right continues today.

In 1919, the Walker River Irrigation District (WRID) was formed to oversee Walker River water allocations in California and Nevada (Shamberger, 1991).

The WRID covers 235,000 acres, of which 80,000 acres are irrigated, and, along with the federal water master, manages the delivery of irrigation rights to 200 miles of drainage ditches in Mason and Smith Valleys (Horton, 1996). Between 1922 and 1934, the district constructed the Bridgeport and Topaz reservoirs to store snowmelt runoff and flood waters to extend the irrigation season and expand the irrigated acreage to areas outside the Walker River Decree (Horton, 1996). Storage rights to these reservoirs are managed solely by WRID. Similarly, the U.S. Bureau of Indian Affairs constructed Weber Reservoir in the 1930s to capture surplus flows to irrigate the Walker River Paiute Tribe's reservation lands (Tiller, 2006). Because the Walker River was fully allocated through both decree and storage rights, flows to Walker Lake began to drop dramatically, and lake levels began a steep downward trend.

Additionally, in the early 1960s, the Nevada State Engineer permitted primary and supplemental groundwater rights to be developed in both Smith and Mason Valleys. The primary groundwater rights were intended to allow for an increase in irrigated acreage or for other uses besides irrigation, such as mining. Supplemental groundwater rights "supplement" surface water rights and may only be used when preexisting surface water rights are insufficient to meet the full irrigation demand due to low river flows. A substantial increase in groundwater pumping in the 1970s and 1980s, coupled with more variable water supply, further decreased surface flows of the Walker River and subsequently groundwater reserves.

Together, surface and groundwater development has drastically impacted Walker Lake, resulting in a nearly 160-foot decline in lake levels and an increase in total dissolved solids concentrations from 2,500 to 25,000 mg/L between 1882 and 2010 (Niswonger et al., 2014). These changes have threatened the lake's freshwater ecosystem and the survival of the Lahontan cutthroat trout, a threatened species under the Endangered Species Act that no longer appears in Walker Lake, in addition to a variety of other fish species and migratory birds that use the lake as a stopover (Niswonger et al., 2014).

Water Rights Acquisition and Temporary Storage Leasing Programs

In the interest of restoring Walker Lake, a nongovernmental organization called the National Fish and Wildlife Foundation (NFWF) utilized federal funds provided through the Desert Terminal Lakes Program (Public Law 109-103, 2005) to create the Walker Basin Restoration Program. Through this program, NFWF helped develop two market-based mechanisms to facilitate both permanent acquisitions and temporary trading of water rights.

The first of these programs facilitates permanent water rights acquisitions for the purposes of environmental restoration at Walker Lake (Kendy et al., 2018). In 2015, NFWF established a nonprofit called the Walker Basin

Conservancy (WBC) to administer this program. Using funding provided by U.S. Congress in 2009 (Public Law 111–85), the WBC has acquired approximately 47 percent of the water rights determined necessary to restore and maintain Walker Lake by purchasing agricultural lands and appurtenant water rights, including decree, storage, and groundwater rights, from willing sellers (Walker Basin Conservancy, 2019). Although permitted in 2014, the first surface rights transferred to Walker Lake were not delivered until the 2019 irrigation season. In order to meet the program's long-term restoration goals, the WBC would need to acquire an estimated 20 percent of all Walker River Basin surface water from willing sellers (Walker Basin Conservancy, 2020).

To help mitigate the negative economic and environmental impacts of removing water rights from irrigated agricultural lands in the Walker River Basin, the WBC has leased some of its acquired primary groundwater rights to facilitate the transition of irrigated grass and alfalfa hay production to vegetable crops, as groundwater is considered a cleaner and more reliable source of irrigation water—particularly for organic produce crops. To date, one established vegetable grower in the basin has purchased WBC's acquired groundwater rights to expand its existing operations (Kendy et al., 2018). The WBC also conducts significant stewardship efforts throughout the basin on their acquired agricultural land and has relinquished most of the supplemental groundwater rights it has acquired back to the state.

The second market-based program is the Stored Water Program (SWP), which was developed through an agreement between NFWF and WRID, who administers the program. Designed as a three-year demonstration program to assess feasibility and user interest, the SWP facilitates temporary leases and transfers of storage water rights in the system's two major reservoirs to increase environmental flows to Walker Lake (California Water Resources Control Board, 2020). Through the SWP, farmers with storage water rights can lease a specified quantity of their stored water in exchange for a per-acre-foot payment amount that is held constant for all storage water enrolled in the program in a given year (Walker River Irrigation District, 2020b). Stored water is then conveyed to Walker Lake through WRID's delivery system. Annual storage water leases may be renewed, subject to WRID board approval, and the program must be approved on an annual basis by the Nevada State Engineer and the California Water Resources Control Board (Walker River Irrigation District, 2020a).

Historically, farmers with storage rights in the basin's two reservoirs have always been able to lease and temporarily transfer stored water among themselves within the irrigation district, subject to WRID governing board approvals that occur during regular monthly meetings. This third water market mechanism has long been in place, providing agricultural users with the means to adapt their water demand annually and is particularly useful in low-runoff years. The SWP, on the other hand, was first administered in 2019, a wet year with well above-normal flows[5] (MBK Engineers, 2019). That year, WRID released approximately 17,600 acre-feet of leased storage rights to Walker Lake, of which an estimated 13,500 acre-feet reached the lake after conveyance losses. The viability of the SWP to enhance environmental flows will be tested in drier years when water supply is significantly below normal.

DISCUSSION AND CONCLUSIONS: WATER MARKETS AS A CLIMATE CHANGE ADAPTATION STRATEGY?

We now return to the overarching question posed at the beginning of this chapter: *How can water markets be used to adapt to hydroclimatic changes in semi-arid, snow-fed river basins?* Part 1 described the water governance structure of the western United States and considered how water markets are currently integrated into this structure. Part 2 described two western river basins that are implementing water market mechanisms. Our goal in this final section is to speculate on the opportunities and challenges associated with water markets as a climate change adaptation strategy in semi-arid, snow-fed river basins by drawing lessons from the cases.

Opportunities for Water Markets

First, there is evidence that the market mechanisms in both cases promote greater water management flexibility under hydrologic uncertainty. For example, in the South Platte River Basin, supplemental C-BT Project water is allocated to users in allotments, or equal shares of the total annual diversion quota, rather than in set amounts of acre-feet of water. Using this system, Northern Water can adjust water allocations under variable water supply conditions by reducing the quantity of water represented by each allotment. Compared to prior appropriation, which would curtail junior users until senior users' rights could be fulfilled, allotments distribute risk equally among allottees in low water years and provide clearer information about the types of water delivery reductions each allottee can expect. This water allocation approach essentially builds in a mechanism for rapid adaptation to variable water supplies. Furthermore, in addition to being able to permanently sell their allotments, C-BT Project allottees can lease portions of their allotments on an annual basis, temporarily store portions of water in Project reservoirs, and even purchase additional annual storage from willing sellers. These trading and transfer options provide water rights holders with enhanced flexibility to make decisions about their water demand in a given year based on a number of factors, including the likelihood their primary water right is fulfilled and the percentage of supplemental C-BT Project water they will receive. In the Walker River Basin, the SWP, at least as it functioned in its first year of operation in 2019, provides similar flexibility to agricultural producers who want to lease a portion or all of their storage water rights in a given year after they have assessed other variables that may affect their decisions about water use.

Both the C-BT allotment system and the SWP also provide an incentive for producers to conserve water: if a farmer conserves some quantity of their water—for example, by changing irrigation technologies or cropping patterns, or by planting a lower-water-use crop on a portion of their land—they have the option to lease the remainder of their conserved stored water to another user on an annual basis. As has always been the case with storage water in the Walker River Basin, farmers can also lease water to other farmers as long as it is approved by WRID, but the SWP now provides an option to temporarily lease water for environmental instream flows, in addition to the option for a permanent water right sale and transfer through the WBC's program. As argued by Anderson et al. (2019), Colby & Isaaks (2018), Culp et al. (2014), Libecap (2018), and others, these types of market mechanisms provide risk management options for different water users to adapt to increasingly uncertain water supplies and can result in enhanced water supply for both agricultural users and the environment.

These case studies also demonstrate the importance of utilizing multiple, interacting market mechanisms to maximize flexibility, which is increasingly necessary for adapting to climate change-induced water supply-and-demand changes. Temporary leases of water or storage space, in particular, provide new options for water rights holders who do not want to permanently sell their water right assets and can increase the robustness of a water market. In addition to augmenting water supplies for potentially drought-stricken cities (Culp et al., 2014), leases also provide the opportunity for agricultural water rights holders to experiment with water trading, including valuation and price negotiation, in order to learn what works best for them before permanently altering their water rights ownership—which may ultimately be a way to avoid yearly administrative changes to water rights that accrue transactions costs (Dilling et al., 2019). Such costs can also be reduced when multiple market mechanisms are administered by a single entity that can provide local oversight and facilitate transparent transactions, such as a conservancy district as seen in the case of the C-BT Project (Colby & Isaaks, 2018).

Challenges for Water Markets

The case studies also revealed a variety of challenges to realizing effective water markets that may undermine their viability as a large-scale climate adaptation solution. Perhaps most obviously, the water markets studied here are highly reliant on trading and transfers of water rights held by agricultural users. This may potentially lead to an increase in the drying up of agricultural lands, especially during prolonged drought periods. While willing sellers are entitled to sell their land and/or water rights, the effects of "buy and dry" can ripple through agricultural communities, creating a variety of negative impacts. Not only does "buy and dry" externalize the costs of growth from urban centers to less-economically diverse rural areas, but it may also introduce new environmental challenges to these areas such as fugitive dust due to lack of irrigated ground cover, invasive weed infestations on previously cultivated lands, and subsequent degradation of soil (Dilling et al., 2019; Squillace, 2013). Critically, the WBC's stewardship activities that seek to maintain the ecological health of lands previously used for agricultural production provide a model for combating this phenomenon elsewhere.

A related challenge is that conserving water in the agriculture sector to lease or sell can be more difficult in practice than it appears. For example, switching to less-water-intensive and/or potentially higher-value crops does not guarantee increased profits (Davitt, 2011) since it often requires farmers to learn new cultivation practices, purchase new equipment, dig agricultural wells and incur groundwater pumping costs, incur higher labor costs, expand their water rights portfolio, and secure advance contracts for their products—most of which require a large amount of upfront operating capital. Additionally, some irrigated crops that have traditionally been considered "water-intensive" have greater actual values due to the ecosystem services they provide or their key role in global supply chains, which may reduce farmers' desire to change crops. For example, alfalfa hay provides ecosystem services such as stabilizing soils and depositing nitrogen and is also a necessary input to a variety of high-value industries like meat production, dairy production, and equine sports (Putnam et al., 2001). Because water remains a relatively low-cost agricultural input in the western United States, the costs involved with switching crops or investing in new irrigation technologies may disincentivize changes to agricultural practices, even when willing buyers or lessees are present (Dilling et al., 2019).

Additionally, many water markets only include trading and transfer of surface water supplies, despite surface water and groundwater often being interconnected, both hydrologically and increasingly legally through conjunctive water management and policies (T. L. Anderson et al., 2012). Water markets that reduce surface water use but simultaneously increase agricultural producers' reliance on groundwater reserves essentially ignore the realities of the physical interconnection between these two water sources and overall water scarcity in the western United States.

Moving Forward with Water Markets

It is evident that while water markets can introduce flexibility into water governance systems characterized by clear water rights, markets alone cannot solve all climate adaptation challenges in the semi-arid, snow-fed western United States. The impacts of climate change-driven water supply volatility on the already delicately balanced water budgets of this region highlight the need for a variety of technological and institutional innovations around water management to occur in tandem with the development of water markets.

On the technological side, more extensive hydrologic modeling, conveyance infrastructure, and irrigation technology could help to conserve additional water supplies for when they are needed the most. Developing and using seasonal, rather than long-term, water forecasts, for example, could inform modest, yet, rapid improvements in agricultural water-use efficiency at each decision point. Improved groundwater data and analysis can also help to reveal long-term usage trends that can identify non-sustainable levels or overdraft, which can improve groundwater management practices (Davitt, 2011). In this vein, Ohja (2018), Culp et al. (2014), Anderson et al. (2012), and others maintain that the proper management of groundwater is critical to overall water-use planning in the context of climate adaptation, including the development and use of highly functional water markets.

Developing stronger institutional arrangements for managing water that enhance information provision and reduce the costs of cooperative and integrated water management are also necessary (Garrick et al., 2018). In the western United States, this includes leadership by state and federal agencies, such as state divisions of water resources, U.S. Bureau of Reclamation and U.S. Army Corp of Engineers, "to promote interstate and interagency cooperation in water management, as well as to coordinate essential state-level gathering of data on water supplies and water use" (Culp et al., 2014, p. 2). Colby and Isaaks (2018) argue that having more publicly available data on water prices, volumes, uses, and changes in seasonal patterns of water use is also necessary to increase broad participation in water markets and to better understand the effects of water markets on different water use sectors.

Other institutional innovations might include the development of water policies in the western United States and other arid areas that reflect the reality of increasingly variable water supplies, the changing marginal value of competing water uses, and the impact of hydrologic changes for other economic sectors. For instance, Tarlock (2018) calls for the development of a "coherent food and water policy in the name of food security" (p. 17), which could help to address negative externalities associated with some market mechanisms, such as "buy-and-dry" practices. Other water management policies change the way infrastructure is operated. For example, changing reservoir operations to allow flexibility for earlier storage and release timing as snowmelt and runoff begin earlier could help to maximize the use of scarce and variable water supplies in semi-arid, snow-fed basins (Sterle et al., 2020),

In sum, water markets present a range of opportunities for adapting to climate change impacts on the water resources sector of the western United States, particularly because most states in the region administer water rights as private property rights. Moreover, markets can take on different shapes and sizes to best fit local conditions, as was demonstrated through the case studies of the water markets in the South Platte and Walker River basins, which provide some common lessons but use a variety of different market mechanisms. That said, policy-makers and water managers should be cautious about an over-reliance on water markets as a singular climate adaptation strategy, especially given the variety of challenges that confront the development of highly functional water markets. Developing water markets in tandem with the technological and institutional innovations described above can further enhance their viability and contribution as a climate change adaptation measure in the western United States and in similar contexts around the world.

NOTES

1. In 1878, geologist and explorer John Wesley Powell conducted extensive land surveys to support the idea that the North American continent is divided into the arid western region and the humid eastern region, separated by a precipitation gradient, at the 100th meridian or longitudinal line. Powell intended his concept to guide the U.S. government in developing the western region within its climatic and hydrologic constraints (Seager et al., 2018). In the United States, the 100th meridian runs through North Dakota and South Dakota, Nebraska, Kansas, Oklahoma, and Texas. In this chapter, we refer to the American West as inclusive of western portions of these states, in addition to Montana, Idaho, Wyoming, Colorado, New Mexico, Arizona, Utah, Nevada, California, Oregon, and Washington.
2. An acre-foot refers to the amount of water needed to cover one acre of land (43,560 square feet) in one foot (12 inches) of water; an acre-foot of water is about 325,851 gallons.
3. The Walker River Decree succeeded the Rickey Decree of 1919, which created an initial surface water allocation system in the basin that included a state-law-based water right for the Walker River Paiute Reservation. This arrangement was later challenged, resulting in a federal reserved water right for the tribe (Puglielli, 2019).
4. "Federal Indian reserved water rights" originate in the Winters Doctrine, based on a U.S. Supreme Court case decision (*Winters v. United States*, 207 U.S. 564 [1908]), whereby the U.S. government must reserve water rights for reservation lands to fulfill the purpose of reservation lands—to become self-reliant through irrigated

agriculture (Cosens & Royster, 2012). Such water rights are reserved by the year in which the U.S. government established a reservation by treaty, executive order, or statute, or by "time immemorial" for aboriginal uses. Since the majority of reservations in the western United States were established in the latter nineteenth century, the Winters Doctrine reserves some of the earliest established water rights. However, many tribes are still working to quantify and access their reserved water rights (Colby et al., 2005; Tarlock, 2010).

5. A water year refers to a twelve-month time period (October 1 to September 30) during which annual precipitation totals are measured. A normal water year refers to a year in which the annual snowmelt runoff volume is within one standard deviation of the mean annual runoff volume for the period of record. Similarly, a wet water or dry water year refers to a year in which the annual runoff volume is greater than or lesser than one standard deviation of the mean annual runoff volume for the period of record (U.S. Geologic Survey [USGS], 2016).

REFERENCES

Adler, J. H. (2012). Water Rights, Markets, and Changing Ecological Conditions. *Environmental Law*, *42*(1), 93–113.

Akhter, M., & Ormerod, K. J. (2015). The irrigation technozone: State power, expertise, and agrarian development in the U.S. West and British Punjab, 1880–1920. *Geoforum*, *60*, 123–132. https://doi.org/10.1016/j.geoforum.2015.01.012

Anderson, S. E., Anderson, T. L., Hill, A. C., Kahn, M. E., Kunreuther, H., Libecap, G. D., Mantripragada, H., Mérel, P., Plantinga, A. J., & Kerry Smith, V. (2019). The critical role of markets in climate change adaptation. *Climate Change Economics*, *10*(01), 1950003. https://doi.org/10.1142/S2010007819500039

Anderson, T. L., Scarborough, B., & Watson, L. R. (2012). *Tapping Water Markets* (1st ed.). Routledge. https://www.routledge.com/Tapping-Water-Markets-1st-Edition/Anderson-Scarborough-Watson/p/book/9780203136072

Barnett, T. P., Pierce, D. W., Hidalgo, H. G., Bonfils, C., Santer, B. D., Das, T., Bala, G., Wood, A. W., Nozawa, T., Mirin, A. A., Cayan, D. R., & Dettinger, M. D. (2008). Human-induced changes in the hydrology of the western United States. *Science*, *319*(5866), 1080–1083. https://doi.org/10.1126/science.1152538

Begay, M. (2018). *Walker River Paiute Tribe Climate Adaptation Plan* (WR-22-2018; p. 37).

Boelens, R., & Zwarteveen, M. (2005). Prices and politics in Andean water reforms. *Development and Change*, *36*(4), 735–758. https://doi.org/10.1111/j.0012-155X.2005.00432.x

Brennan, S. (2017). A review of water banking in state legislation of the western United States [Master's Thesis]. Oregon State University.

Brewer, J., Glennon, R., Ker, A., & Libecap, G. (2008). Water markets in the West: Prices, trading, and contractual forms. *Economic Inquiry*, *46*(2), 91–112. https://doi.org/10.1111/j.1465-7295.2007.00072.x

Burness, H. S., & Quirk, J. P. (1979). Appropriative water rights and the efficient allocation of resources. *The American Economic Review, 69*(1), 25–37.

California Department of Water Resources. (1992). *Walker River Atlas*. State of California, Department of Water Resources. http://images.water.nv.gov/images/publications/River%20Chronologies/Walker_River_Atlas.pdf

California Water Resources Control Board. (2020). *Notice of Petitions for Temporary Transfer and Change Filed by Walker River Irrigation District licenses 6000 and 9407 (Applications 2221 and 1389) and Involving Federally Adjudicated Rights Established by the Walker River Decree*. California Water Boards. https://www.waterboards.ca.gov/waterrights/water_issues/programs/applications/transfers_tu_notices/2020/a1389_transfer_notice_2020.pdf

Carey, J. M., & Sunding, D. L. (2001). Emerging markets in water: A comparative institutional analysis of the Central Valley and Colorado-Big Thompson projects. *Natural Resources Journal, 41*(2), 283–328. JSTOR.

Clifford, P., Landry, C., & Larsen-Hayden, A. (2004). *Analysis of Water Banks in the Western States* (No. 04-11–011). Washington Department of Ecology and WestWater Research. https://fortress.wa.gov/ecy/publications/publications/0411011.pdf

Colby, B. G., & Isaaks, R. (2018). Water trading: Innovations, modeling prices, data concerns. *Journal of Contemporary Water Research & Education, 165*(1), 76–88. https://doi.org/10.1111/j.1936-704X.2018.03295.x

Colby, B. G., Thorson, J. E., & Sarah Britton. (2005). *Negotiating tribal water rights: Fulfilling promises in the arid West*. The University of Arizona Press.

Colorado Water Conservation Board. (2011). *SWSI 2010 Mission Statement, Key Findings, and Recommendations*.

Cosens, B., & Royster, J. V. (Eds.). (2012). *The future of Indian and federal reserved water rights: The winters centennial*. University of New Mexico Press.

Culp, P. W., Glennon, R., & Libecap, G. (2014). *Shopping for Water: How the Market Can Mitigate Water Shortages in the American West*. The Hamilton Project and the Stanford Woods Institute for the Environment. https://www.brookings.edu/wp-content/uploads/2016/06/market_mitigate_water_shortage_in_west_glennon.pdf

Davitt, A. (2011). *Climate Variability and Drought in the South Platte River Basin*. City University of New York. https://academicworks.cuny.edu/cgi/viewcontent.cgi?referer=https://www.google.com/&httpsredir=1&article=1035&context=cc_etds_theses

Diffenbaugh, N. S., Scherer, M., & Ashfaq, M. (2013). Response of snow-dependent hydrologic extremes to continued global warming. *Nature Climate Change, 3*, 379–384. https://doi.org/10.1038/nclimate1732

Dilling, L., Berggren, J., Henderson, J., & Kenney, D. (2019). Savior of rural landscapes or Solomon's choice? Colorado's experiment with alternative transfer methods for water (ATMs). *Water Security, 6*, 100027. https://doi.org/10.1016/j.wasec.2019.100027

Ganoe, J. T. (1937). The Desert Land Act in Operation, 1877–1891. *Agricultural History, 11*(2), 142–157.

Garrick, D. E., Hernández-Mora, N., & O'Donnell, E. (2018). Water markets in federal countries: Comparing coordination institutions in Australia, Spain and the

Western USA. *Regional Environmental Change*, *18*(6), 1593–1606. https://doi.org/10.1007/s10113-018-1320-z

Ghosh, S. (2019). Droughts and water trading in the western United States: Recent economic evidence. *International Journal of Water Resources Development*, *35*(1), 145–159. https://doi.org/10.1080/07900627.2017.1411252

Glennon, R. (2009). *Unquenchable: America's Water Crisis and What To Do About It*. Island Press.

Grafton, R. Q., Libecap, G., McGlennon, S., Landry, C., & O'Brien, B. (2011). An integrated assessment of water markets: A cross-country comparison. *Review of Environmental Economics and Policy*, *5*(2), 219–239. https://doi.org/10.1093/reep/rer002

Guy, R. K., Boykin, K. G., Kepner, W. G., & McCarthy, J. M. (2011). *South Platte River Basin Data Browser*. Environmental Protection Agency (EPA). https://cfpub.epa.gov/si/si_public_record_report.cfm?Lab=NERL&dirEntryId=240251

Holley, C., & Sinclair, D. (2018). Water markets and regulation: Implementation, successes and limitations. In C. Holley & D. Sinclair (Eds.) *Reforming water law and governance: From stagnation to innovation in Australia* (pp. 141–168). Springer. https://doi.org/10.1007/978-981-10-8977-0_7

Horton, G. (1996). *Walker River Chronology: A Chronological History of the Walker River and Related Water Issues*. Nevada Division of Environmental Protection, Bureau of Water Quality Planning.

Howe, C. W. (2011). *The Efficient Water Market of the Northern Colorado Water Conservancy District: Colorado, USA* (D6.1-IBE Review Reports). EPI Water. https://www.issuelab.org/resource/the-efficient-water-market-of-the-northern-colorado-water-conservancy-district-colorado-usa.html

Howitt, R. E., Medellín-Azuara, J., MacEwan, D., & Lund, J. R. (2012). Calibrating disaggregate economic models of agricultural production and water management. *Environmental Modelling & Software*, *38*, 244–258. https://doi.org/10.1016/j.envsoft.2012.06.013

Intergovernmental Panel on Climate Change. (2015). *Climate Change 2014: Synthesis Report* (p. 151). Intergovernmental Panel on Climate Change. https://www.ipcc.ch/site/assets/uploads/2018/02/SYR_AR5_FINAL_full.pdf

Jones, L., & Colby, B. (2010). Weather, climate, and environmental water transactions. *Weather, Climate, and Society*, *2*(3), 210–223. https://doi.org/10.1175/2010WCAS1028.1

Kendy, E., Aylward, B., Ziemer, L. S., Richter, B. D., Colby, B. G., Grantham, T. E., Sanchez, L., Dicharry, W. B., Powell, E. M., Martin, S., Culp, P. W., Szeptycki, L. F., & Kappel, C. V. (2018). Water transactions for streamflow restoration, water supply reliability, and rural economic vitality in the western United States. *JAWRA Journal of the American Water Resources Association*, *54*(2), 487–504. https://doi.org/10.1111/1752-1688.12619

Lall, U., Johnson, T., Colohan, P., Aghakouchak, A., Arumugam, S., Brown, C., Mccabe, G. J., & Pulwarty, R. S. (2018). *Chapter 3: Water* (Volume II; Impacts, risks, and adaptation in the United States: The Fourth National Climate

Assessment, pp. 145–173). U.S. Global Change Research Program. https://doi.org/10.7930/NCA4.2018.CH3

Landstrom, K. S. (1954). Reclamation under the Desert-Land Act. *Journal of Farm Economics*, *36*(3), 500–508.

Leonard, B., & Libecap, G. D. (2015). Endogenous first-possession property rights in open-access resources. *Iowa Law Review*, *100*, 2457–2478.

Leonard, B., & Libecap, G. D. (2019). Collective action by contract: Prior appropriation and the development of irrigation in the western United States. *Journal of Law and Economics*, *62*, 67–115.

Libecap, G. D. (2011). Institutional path dependence in climate adaptation: Coman's "Some Unsettled Problems of Irrigation." *The American Economic Review*, *101*(1), 64–80.

Libecap, G. D. (2018). Policy note: Water markets as adaptation to climate change in the western United States. *Water Economics and Policy*, *4*(3). https://doi.org/10.1142/S2382624X18710030

Lopes, T. J., & Allander, K. K. (2009). *Water Budgets of the Walker River Basin and Walker Lake, California and Nevada* (Scientific Investigations Report No. 2009–5157; Scientific investigations report, pp. 1–44). USGS; Bureau of Reclamation.

Lopes, T. J., & Smith, J. L. (2007). *Bathymetry of Walker Lake, West-Central Nevada* (Scientific Investigations Report No. 2007–5012). USGS; Bureau of Reclamation. https://pubs.usgs.gov/sir/2007/5012/pdf/sir20075012.pdf

Lounsberry, S. (2019). Front Range farmers look to cities to lease water as prices surge. *Water Education Colorado*. https://www.watereducationcolorado.org/fresh-water-news/front-range-farmers-look-to-cities-to-lease-water-as-prices-surge/

Lueck, D. (1995). The rule of first possession and the design of the law. *The Journal of Law and Economics*, *38*, 393–436.

Maas, A., Dozier, A., Manning, D. T., & Goemans, C. (2017). Water storage in a changing environment: The impact of allocation institutions on value. *Water Resources Research*, *53*(1), 672–687. https://doi.org/10.1002/2016WR019239

Mankin, J. S., Viviroli, D., Singh, D., Hoekstra, A. Y., & Diffenbaugh, N. S. (2015). The potential for snow to supply human water demand in the present and future. *Environmental Research Letters*, *10*(11), 114016. https://doi.org/10.1088/1748-9326/10/11/114016

MBK Engineers. (2019). *WRID 2019 Release Plan*. http://www.wrid.us/WRID/News

Milly, P. C. D., & Dunne, K. A. (2020). Colorado River flow dwindles as warming-driven loss of reflective snow energizes evaporation. *Science*, *367*(6483), 1252–1255.

Murphy, J. J., Dinar, A., Howitt, R. E., Rassenti, S. J., Smith, V. L., & Weinberg, M. (2009). The design of water markets when instream flows have value. *Journal of Environmental Management*, *90*(2), 1089–1096. https://doi.org/10.1016/j.jenvman.2008.04.001

National Research Council. (2005). *Endangered and Threatened Species of the Platte River* (p. 10978). National Academies Press. https://doi.org/10.17226/10978

Nevada Division of Water Planning. (1999). *Nevada State Water Plan: Part 1-background and Resource Assessment, Section 8 Glossary on Selected Water-Related Decrees, Agreements and Operating Criteria*. Nevada Division of Water Planning. http://water.nv.gov/programs/planning/stateplan/documents/pt1-sec8.pdf

Niswonger, R. G., Allander, K. K., & Jeton, A. E. (2014). Collaborative modelling and integrated decision support system analysis of a developed terminal lake basin. *Journal of Hydrology*, *517*, 521–537. https://doi.org/10.1016/j.jhydrol.2014.05.043

Northern Colorado Water Conservancy District. (2020a). *Allottee Documents*. Northern Water. https://www.northernwater.org/sf/allottee-information/allottee-documents

Northern Colorado Water Conservancy District. (2020b). *Northern Water C-BT Project*. Northern water. https://www.northernwater.org/WaterProjects/C-BTProject.aspx

Northern Colorado Water Conservancy District. (2020c). *Northern Water C-BT Quota*. Northern water. https://www.northernwater.org/AllotteeInformation/C-BTQuota.aspx

O'Donnell, M., & Colby, Dr. B. (2010). *Water Banks: A Tool for Enhancing Water Supply Reliability*. The University of Arizona, Department of Agricultural and Resource Economics. https://cals.arizona.edu/arec/sites/cals.arizona.edu.arec/files/publications/ewsr-Banks-final-5-12-10.pdf

Ojha, C., Shirzaei, M., Werth, S., Argus, D. F., & Farr, T. G. (2018). Sustained groundwater loss in California's Central Valley exacerbated by intense drought periods. *Water Resources Research*, *54*, 4449–4460. https://doi-org.unr.idm.oclc.org/10.1029/2017WR022250

Puglielli, G. (2019, February 18). *United States v. Walker River Irrigation District*. University of Denver Water Law Review at the Sturm College of Law. http://duwaterlawreview.com/united-states-v-walker-river-irrigation-district/

Putnam, D., Russelle, M., Orloff, S., Kuhn, J., Fitzhugh, L., Godfrey, L., Kiess, A., & Long, R. (2001). *Alfalfa, Wildlife and the Environment: The Importance and Benefits of Alfalfa in the 21st Century* (pp. 1–23). California Alfalfa and Forage Association. http://agric.ucdavis.edu/files/242006.pdf

Regonda, S. K., Rajagopalan, B., Clark, M., & Pitlick, J. (2005). Seasonal cycle shifts in hydroclimatology over the western United States. *Journal of Climate*, *18*(2), 372–384. https://doi.org/10.1175/JCLI-3272.1

Schilling, K. (2018). Addressing the prior appropriation doctrine in the shadow of climate change and the Paris Climate Agreement. *Seattle Journal of Environmental Law*, *8*(1), 97-119.

Schutz, J. R. (2012). Why the Western United States' prior appropriation water rights system should weather climate variability. *Water International*, *37*(6), 700–707. https://doi.org/10.1080/02508060.2012.726544

Schwabe, K., Nemati, M., Landry, C., & Zimmerman, G. (2020). Water markets in the western United States: Trends and opportunities. *Water*, *12*, 1–15. https://doi.org/10.3390/w12010233

Seager, R., Lis, N., Feldman, J., Ting, M., Williams, A. P., Nakamura, J., Liu, H., & Henderson, N. (2018). Whither the 100th meridian? The once and future physical and human geography of America's arid–humid divide. Part I: The story so far. *Earth Interactions*, *22*(5), 1–22. https://doi.org/10.1175/EI-D-17-0011.1

Shamberger, H. A. (1991). *Evolution of Nevada's Water Laws, as Related to the Development and Evaluation of the State's Water Resources, Form 1866 to About 1960* (Water-Resources Bulletin No. 46). USGS; Nevada Division of Water Resources. http://images.water.nv.gov/images/publications/water%20resources%20bulletins/Bulletin46.pdf

Smith, S. M. (2019). Instream flow rights within the prior appropriation doctrine: Insights from Colorado. *Natural Resources Journal*, *59*(1), 181–213.

South Platte Basin Roundtable. (2019). *Water Facts: South Platte River Basin* [Fact Sheet]. https://southplattebasin.com/wp-content/uploads/2019/05/South-Platte-River-Basin-Infographic-2019-5-print.pdf

South Platte Basin Roundtable/Metro Basin Roundtable. (2015). *South Platte Basin Implementation Plan.* https://www.colorado.gov/pacific/sites/default/files/SouthPlatteBasinImplementationPlan-04172015.pdf

South Platte Coalition for Urban River Evaluation. (2015). *Total Maximum Daily Load Assessment: Upper South Platte Segment 15* (p. 72). Colorado Department of Public Health and Environment, Water Quality Control Division.

Squillace, M. (2013). Water transfers for a changing climate. *Natural Resources Journal*, *53*(1), 55–116. JSTOR.

Statewide Water Supply Initiative (SWSI). (2011). *SWSI 2010 Mission Statement, Key Findings, and Recommendations.* Colorado Water Conservation Board; Colorado Department of Natural Resources.

Stegner, W. (1954). *Beyond the Hundredth Meridian—John Wesley Powell and the Second Opening of the West.* Penguin Books.

Sterle, K., Jose, L., Coors, S., Singletary, L., Pohll, G., & Rajagopal, S. (2020). Collaboratively modeling reservoir reoperation to adapt to earlier snowmelt runoff. *Journal of Water Resources Planning and Management*, *146*(1), 05019021. https://doi.org/10.1061/(ASCE)WR.1943-5452.0001136

Tarlock, A. D. (2002). The future of prior appropriation in the West. *Natural Resources Journal*, *41*, 769–793.

Tarlock, A. D. (2010). Tribal justice and property rights: The evolution of Winters v. United States. *Natural Resources Journal*, *50*(2), 471–499.

Tarlock, A. D. (2018). Western water law and the challenge of climate disruption. *Environmental Law*, *48*(1), 1–27.

Tiller, V. E. V. (Ed.). (2006). *Tiller's guide to Indian country: Economic PROFILES of American Indian reservations.* Bow Arrow Publishing Co.

Tracy, J. C. (2004). Environmental impacts from water management in a closed basin: Walker Lake, Nevada. *Southwest Hydrology*, 14–17.

Udall, B., & Overpeck, J. (2017). The twenty-first century Colorado River hot drought and implications for the future. *Water Resources Research*, *53*(3), 2404–2418. https://doi.org/10.1002/2016WR019638

U.S. Bureau of Reclamation. (2010). *Walker River Basin Acquisition Program: Revised Draft Environmental Impact Statement.* https://www.usbr.gov/mp/nepa/nepa_project_details.php?Project_ID=2810

U.S. Geologic Survey (USGS). (2016). *Explanations for the National Water Conditions*. Water resources of the United States. https://water.usgs.gov/nwc/explain_data.html

U.S. Global Change Research Program. (2018). *Impacts, Risks, and Adaptation in the United States: The Fourth National Climate Assessment, Volume II*. U.S. Global Change Research Program. https://doi.org/10.7930/NCA4.2018

Walker Basin Conservancy. (2019). *Walker Basin Conservancy Completes New Water Acquisition for the Restoration of Walker Lake.*

Walker Basin Conservancy. (2020). *Water Conservation* [Walker Basin Conservancy]. https://www.walkerbasin.org/water-conservation

Walker River Irrigation District. (2020a). *January 7, 2020 WRID board Minutes.* http://www.wrid.us/WRID/Meetings

Walker River Irrigation District. (2020b). *Walker River Irrigation District.* WRID. http://www.wrid.us/

Water Education Colorado. (2019). *Citizen's Guide to Where Your Water Comes from (2nd Edition).* https://www.watereducationcolorado.org/publications-and-radio/citizen-guides/citizens-guide-to-where-your-water-comes-from/

Wattage, D. P., Smith, A., Pitts, D. C., McDonald, A., & Kay, D. (2000). Integrating environmental impact, contingent valuation and cost-benefit analysis: Empirical evidence for an alternative perspective. *Impact Assessment and Project Appraisal, 18*(1), 5–14. https://doi.org/10.3152/147154600781767600

Woodhouse, C. A., Lukas, J. J., Morino, K., Meko, D. M., & Kirschboeck, K. K. (2016). Using the Past to Plan for the Future—The Value of Paleoclimate Reconstructions for Water Resource Planning. In Kathleen A. Miller, Alan F. Hamlet, Douglas S. Kenney, Kelly T. Redmond. *Water Policy and Planning in a Variable and Changing Climate* (pp. 161–182). CRC Press. https://doi.org/10.1201/b19534

Chapter 11

Do Environmental Policies Enhance Environmental Quality?

An Examination of Policy Instruments and Outcomes

Emilia Barreto Carvalho

INTRODUCTION

In a contemporary global society, almost any imagined good can be produced at scale—from food necessary to feed the 7.8 billion people on the planet to mobile devices that connect about 5 billion of them (Silver, 2019). However, in addition to all these goods, unwanted consequences are also produced, such as industrial residues and waste are released into the air, water, and land. Without treatment, the byproduct of industrial activity pollutes and degrades the environment. For instance, the Intergovernmental Panel on Climate Change (IPCC) has indicated that climate change is a direct consequence of anthropogenic actions (IPCC Report, 2019).

The public, scientific community, businesspeople, and governments have attempted to change the course of climate crisis. In this chapter, I examine the role of governments in addressing the problem. I analyze the extent to which government-led environmental policies have promoted environmental quality. I argue that under international and domestic pressures, governments have designed and implemented environmental policies that reduce the impact of the unwanted consequences of production.

Countries such as Australia, Canada, the Netherlands, and the UK have implemented some of the most stringent environmental policies, while others such as Brazil, Indonesia, and Russia have established fairly lax ones (OECD, 2017). As it is characteristic of public policies, environmental policies present costs and benefits. The costs are associated with increased production costs and unemployment in specific industrial sectors, as discussed in the following pages. On the other hand, the benefits are connected to the mitigation of unwanted consequences and the improvement of environmental quality. This chapter focuses on the potential benefits of environmental policies. I investigate the extent to which they have improved environmental quality between 1990 and 2015. More specifically, I explore the impact of these policies on greenhouse gas (GHG) emissions, the chief contributor to climate change. To the best of my knowledge, no analysis of environmental policy outcomes has examined the impact of environmental policies across twenty countries and twenty-five years. Here the dependent variable is annual GHG emissions (OECD, 2020a) and the independent variable is OECD's environmental policy index which measures the degree to which environmental policies put an explicit or implicit price on environmental externalities (OECD, 2017). The use of the OECD index promises more reliable calculations of environmental policy outcomes than, for example, the number of environmental legal acts adopted by a country (see Knill et al., 2010).

Wide-ranging and far-reaching analyses of environmental policy outcomes, as the one performed here, are relevant in handling a challenge as great as climate change. They may indicate the impact of environmental policies on environmental quality and thus contribute to the (re)design of policies that effectively increase environmental quality. The analyses performed in this chapter suggest that although stringent environmental policies are associated with an incremental reduction of GHG emissions, such reduction has not changed the trend of mean GHG emissions.

Here I would like to explain how this chapter is structured. In the first section, I discuss the production of environmental externalities. Next, in the second part, I analyze different types of environmental regulations and how each type addresses externalities. Then, I investigate the impact of environmental policies on GHG emissions. The last section presents the results of statistical regressions and offers some closing thoughts.

THE UNWANTED CONSEQUENCES OF PRODUCTION: ENVIRONMENTAL EXTERNALITIES

Environmental policies are designed to address the unwanted consequences of production. Since policies are most effective when they address the causes of the problems they attempt to solve, in this section, I discuss the causes of the environmental crisis: the unwanted consequences of industrial production. As examples present information to readers and make arguments clearer, I start with one. Imagine, for instance, a regular person in the United States.[1] Suppose that this person wakes up in a heated or cooled house, depending on the season of the year, and drowsily checks their smartphone. After a few minutes swiping and double-clicking, the person gets up, brushes their teeth, takes a shower, and changes into one of the many outfits hanging in their closet. If this person is a woman, she may apply cosmetic products to her face. Once ready, this person grabs the car keys and runs out the door. A latte macchiato with almond milk on their way to work supplies the right dose of energy to start the workday.

In the first hour of the day, this person uses several products such as smartphones, hygiene and cosmetic products packaged in single use plastic containers, synthetic clothes from fast-fashion chain stores with outsourced production chains, and disposable coffee cups or pods, and services such as cooling or heating, internet, water, electricity, and maybe gasoline. This consumption-based lifestyle is the hallmark of contemporary urban society. While I do not discuss the virtues and vices of such lifestyle,[2] its unwanted consequences have a real impact in the world.

The individual in the example consumes several industrialized products. These products are frequently produced, processed or assembled in distant countries with parts from several others, and once ready, they are shipped to stores. In the production process, raw materials, labor (often cheap labor from developing countries), and energy (including energy for transportation) are used. While this process creates cheap goods for mass consumption, it has unwanted consequences. Residues and waste that also result from the industrial production are frequently discarded into the air, water, and soil without treatment. Discharge of contaminated residues and waste causes pollution. For example, as factories burn fossil fuels to generate energy and power machinery, GHG is emitted to the atmosphere and the cumulative release of GHG since the First Industrial Revolution in the late 1900s are considered by the IPCC as the main cause of climate change (IPCC Report, 2019).

Albeit GHG emissions have caused the global temperatures to rise, unwanted consequences of industrial production have been largely neglected, in part, due to skepticism (Jacques & Knox, 2016; Lewandowsky et al., 2015; Smith, 1998), but mostly due to the costs of addressing them. Mitigating the unwanted consequences of industrial production is costly. The environmental costs of industrial production are *not* usually included in the price of goods and end up being paid by society. That is, these environmental costs are divided among many people, and each person pays a fraction of the total price. Therefore, the inclusion of these costs into the price tends to increase the price paid by consumers. In other words, the "new" price shifts the supply-and-demand curves in a market.

In economic terms, the level of industrial output in a market is determined by the supply-and-demand curves. Firms produce goods to meet the needs and wants of consumers. Traditionally, firms incorporate the private costs of production into their prices. Examples of private costs of production are raw materials, labor, rent, utilities, depreciation, and so on. However, firms have not usually included the social costs of production to their prices. The social costs of production include social and environmental-related costs, such as occupational injuries, deforestation, air, water and land pollution, acid rain, and GHG emissions.

As environmental costs are not incorporated into the price of goods, such costs are divided among many individuals. In fact, it is common that the consumers of a good are not the ones most affected by its environmental costs. For example, the Texas Gulf Coast is home to six oil refineries. Oil refined in these facilities fuels the vehicles of millions of consumers all over America. In addition to refining oil, these refineries are the country's largest polluters of benzene, a chemical compound known to cause various forms of cancer. Individuals who live in the vicinity of these plants and breathe the benzene-polluted air pay some high costs. These individuals are more likely to develop forms of cancer associated to exposure to benzene than individuals that live away from the plants (Collins, 2020). That is, the individuals who live near these refineries pay the highest environmental costs, but the benefits of cheap gas are spread in American society. If the environmental costs of production were included in the price of gas, it would be much more expensive.

In times when only the private costs of production are added to a good's price, the market reaches one equilibrium. The inclusion of social and environmental costs shifts this equilibrium. The difference between equilibrium one, which does not consider social costs, and equilibrium two, which adds them to the final price, demonstrates the unwanted consequences of production, commonly referred to as environmental externalities. The price for the same quantity of a good is much higher in equilibrium two. As a result, the demand for this product is lower since fewer consumers are willing or able to pay the full price of the product. As fewer consumers are willing to pay the full price, sales decrease. As sales decrease, production levels also do. Consistently lower sales and reduced production levels tend to generate unemployment, as I discuss in the next section.

ENVIRONMENTAL POLICIES: STRINGENCY AND TYPES OF REGULATIONS

Considering that the inclusion of environmental costs into the price of goods and services may have negative economic consequences, societies all over the world face a dilemma between economic growth and environmental quality. On one hand, the industrial revolutions have promoted the quality of life of billions of individuals. Though the process started in the developed world, it has incorporated developing countries, especially since the 1990s as a result of the liberalization of markets on a global scale. Developing countries' gains with trade liberalization are in the billions of dollars added to GDP growth. As a result of economic integration, billions of people were taken out of poverty and experienced an increased quality of life (Soto, 2000; World Trade Organization, 2018). On the other hand, economic development has caused climate change (Arndt et al., 2012; Dąbrowska et al., 2017; Gonzalez-Perez & Leonard, 2017; IPCC Report, 2019).

Under these complex circumstances, governments are pressured to act by international actors, such as the United Nations (UN), an international organization, and domestic ones, such as the public opinion, environmentalists and labor unions, for example. Gourevitch (1978) and Putnam (1988) have long argued that international pressure compels states to behave in ways that promote the international *status quo*. Contemporary international *status quo* favors a kind of development that promotes both economic growth *and* environmental quality. The secretary-general of the UN, António Guterrez, has summoned world leaders, politicians, businesspeople, the scientific community, and the public to change the course of the environmental crisis (Guterrez, 2018). The European Union has similarly pressured member-states to adopt strict environmental policies.

Nevertheless, domestic actors also influence governments. Opinion surveys suggest that the public in various countries recognize the dilemma posed by economic growth and environmental quality (Capstick, 2013). In democracies such as the United States, the civil society has pressured governments to craft solutions to the environmental crisis without losing sight of economic growth. In these societies, politicians are expected to cater to their bases in order to remain in office. This argument, proposed by Mayhew (2004), is widely recognized in political science and offers a line of reasoning advanced in this chapter. In this sense, public opinion is relevant. Politicians take it as a cue and behave in ways that advances the interests of their support bases (Kingdom, 1993; Zahariadis, 2007). Therefore, opinion surveys are instrumental to politicians' assessment of the interests for the development of public policies that respond to such interests in democratic societies.

Yet, Brechim & Bhandari (2011) have explored the results of public opinion surveys on environmental issues in more than twenty countries and their findings indicate that perceptions regarding the gravity of global warming and the public's willingness to pay the costs of addressing vary greatly within and among countries. The idea that regulations hurt the economy is popular among certain segments of the public and explains why some groups oppose environmental regulations. For example, since environmental regulations are commonly referred to as "job-killing regulations" in the United States (Johnson & Finkel, 2016), many American workers tend to oppose them.

In view of international pressure and domestic factors, governments establish their environmental policies. These policies are composed of different regulatory instruments with varying stringency levels. McGillivray (2004) argues that public policies redistribute costs and benefits among groups in a society. In other words, environmental policies present a tradeoff between environmental quality and economic development. Depending on the composition of a country's economic activities (manufacturing versus agriculture, for example) and on civil society's relationship with the environment (for example, a small island where fishing is the basis of the population's diet versus a group living in the vicinity of an oil refinery which depends on the plant for jobs but, at the same time, is more likely to develop cancer forms associated with the exposure to chemicals released), among other factors, each country establishes a singular environmental policy.

Environmental policy is the body of environmental regulations. That is, the sum of all the different types of regulations establishes a country's environmental policy. The Environmental Policy Stringency (EPS) Index (OECD, 2017) attributes weights to different types of environmental regulations. The sum of the weights results in a stringency measure, and the index describes the EPS level of several countries in the world. The index indicates that as governments attempt to balance international and domestic pressures, the mean stringency level of environmental policies has increased in twenty of the world's largest economies in the period from 1990 to 2018 (Graph 1).

The EPS index classifies environmental policy instruments in two categories: market and nonmarket-based regulations. Market-based regulations have two subcategories: the first establishes a price for environmental degradation. With this, the cost of environmental externalities is calculated and incorporated into the price of goods and services. The costs associated with environmental degradation are transferred from society to consumers through taxes, fees, and tariffs. These types of regulation provide firms with incentives to adjust their production to a new equilibrium. Some common examples of environmental taxes are gas taxes and waste management fees. The revenue from these taxes and fees should be used in the promotion of environmental quality. The same applies to certain tariffs. For example, Feed-In Tariffs (FITs) are paid by electricity consumers to support the build-up of renewable energy capacity. Nevertheless, critics argue that environmental regulations do not affect the production of negative externalities, that is, as long as the market is willing to pay the price of environmental degradation, externalities continue to be produced.

The second subcategory of market-based regulations offers incentives for non-pollution. For example, Deposit and Refund Schemes (DRS), which are different from environmental taxes, attempt to reduce environmental

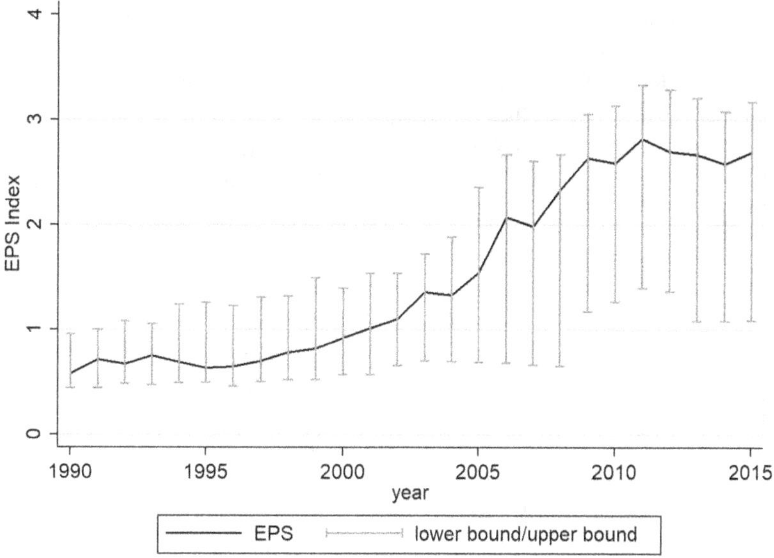

Figure 11.1 Environmental Policy Stringency Index. *Data source*: OECD, 2017.

degradation through incentives for non-pollution. DRS reward conservation and thus promote environmental quality.

Emission trading schemes (ETS) are another example of a regulation that offers incentives for non-pollution. This type of environmental regulation has received attention since the early 2000s, with the implementation of the EU ETS. Such schemes introduce an innovative possibility to dealing with environmental externalities. The theory basing ETS suggests that the lowest cost solution to dealing with environmental externalities may be associated with creating a market for environmental externalities.

As the market for environmental externalities is established through property rights allocation, the price of environmental externalities regulates the production of goods and services. Moreover, as a market becomes more competitive and efficient, manufacturers determine whether to produce or to sell their right to do so. Consequently, only the most efficient manufacturers remain in business, guaranteeing the most production at the lowest cost and maintaining the production of environmental externalities constant. Despite the difficulty to allocate property rights among large groups and the transaction costs associated with it, ETS are considered to produce better environmental and economic outcomes than market-based regulations. Governments have a relevant role in ETS: they allocate property rights, design, implement, and monitor the scheme.

Curiously, the first implemented ETS in the world was the market for sulfur dioxide in the United States in the 1990s. Though extremely successful in reducing sulfur dioxide emissions and remedying the acid rain problem, the market was deregulated a few years after its implementation (Schmalensee & Stavins, 2013). The positive experience with the sulfur dioxide market influenced the design and implementation of the EU ETS (Denny Ellerman & Montero, 2007; Joskow & Schmalensee, 1998; Schmalensee & Stavins, 2013).

The second category in the EPS index refers to nonmarket-based regulations. These types of regulations do not put a price on degradation. Instead, they promote environmental quality through investments in research and development. They include emission standards and subsidies to R&D.

Emission standards impose emission limits to industrial production either by establishing quotas or by specifying a technology to be used. Critics argue that emission standards limit environmental externalities at a high cost: they undermine the supply-and-demand logic of the market and promote higher prices. Despite criticisms, emission standards and subsidies have been successfully used by governments all over the world to phase-out CHC gases which caused the hole in the ozone layer (Mullin, 2002; Smith, 1998; United Nations, 1987). Porter and Linde (Porter & Linde, 1995) argue that regulations which promote innovation, such as emission standards and subsidies, may increase productivity, generate efficiency, and even offset the costs of compliance in the long run. Subsidies to R&D contribute to the development of technologies that use energy sources which burn less or do not burn fossil fuels, promoting environmental quality. In line with Porter and Linde, Harrison et al. (2015) show that emission standards have improved air quality in India with a small impact on productivity.

Harrison et al. (2015) contend that the best environmental policies adopt a vast array of regulatory instruments, adapted to the domestic characteristics of each country. Nevertheless, the question is: what are these policies' environmental outcomes? Do more stringent environmental policies promote environmental quality?

POLICY OUTCOMES

Governments establish environmental policies responding to international and domestic pressures and considering economic and environmental outcomes. The international status quo has pressured governments to fight the environmental crisis, as exemplified by the UN secretary-general speech (Guterrez, 2018). International pressure has been effective in the promotion of different norms and values, including environmental conservation (Barnett & Finnemore, 2004; Elkins et al., 2006; Gourevitch, 1978; Putnam, 1988; Simmons & Elkins, 2004). On the domestic politics realm, representatives of different groups in society push for policies that further their groups' most pressing interests. As representatives promote their base's interests, they increase their chances of winning reelection. This argument is grounded on Mayhew's widely recognized proposition that politicians' number one goal is to remain in office (Mayhew, 2004). For instance, politicians who cater to workers tend to oppose to strict environmental regulations because such regulations are associated with job losses. On the other hand, representatives of disadvantaged populations, such as those located near oil refineries on the Texas Gulf Coast and more likely to develop forms of cancer associated to exposure to benzene (Collins, 2020), tend to support strict regulations since these regulations would contribute to improved health conditions of their support base. The combination of different regulatory instruments, with varying stringency levels becomes a country's environmental policy.

International and domestic pressures combine with unique factors such as economic, social, and cultural characteristics produce singular environmental policies. The OECD has examined the environmental regulations that compose environmental policies and compiled an EPS index. Using the EPS index as a point of departure, I examine the extent to which the stringency of environmental policies in twenty of the world's largest economies has impacted environmental quality in the period between 1990 and 2015. However, environmental quality is a broad concept. It is associated with clean air, water and soil, affordable and clean energy, sustainable cities and communities, and so on. Nevertheless, the focus of this chapter is climate change. Considering that the main component of climate change is GHG emissions, I investigate how environmental policies affect GHG emissions.

The majority of GHG emissions is associated with fossil fuel burning and the EPS index is composed of regulations that either penalizes GHG emissions (such as CO_2, NO_X, SO_X and diesel taxes, emission standards) or incentivizes emission reduction through the development of green technology (trading schemes, FITs[3], DRS,[4] and subsidies to investments on renewable energy, for example). Therefore, the examined environmental regulations have been adopted with the objective of altering the course of climate change through a reduction of GHG emissions. As the number of environmental regulations increases and establish more strict penalties to GHG emissions or further incentivize the development of green technologies, environmental policies become more stringent. The EPS index demonstrates that environmental policies have become more stringent over time (see graph 1). Therefore, considering that the goal of the environmental policies described by the EPS index is GHG emission reduction, I expect that

Hypothesis: As the mean stringency of environmental policies have increased over time, mean GHG emissions have decreased.

RESEARCH METHODS

As I investigate the hypothesis, I analyze annual data for twenty of the largest world economies in the period between 1990 and 2015. The following countries are included in the sample: Australia, Brazil, Canada, China, France, Germany, India, Indonesia, Italy, Japan, Korea, Mexico, Netherlands, Russia, South Africa, Spain, Switzerland, Turkey, United Kingdom, and United States. The sum of these countries' GHG emission levels correspond to about 80 percent of global emissions (OECD, 2020b).

The dependent variable is environmental quality, measured as annual GHG emissions (OECD, 2020).[5] The key independent variable is EPS as measured by the EPS index (OECD, 2017). As mentioned above, the index is composed of market and nonmarket-based environmental regulations.

To increase the reliability of the statistical model, I use several control variables. As economic activities that burn fossil fuels are the main cause of GHG emissions, I include a measure of the annual percentage of GDP per capita growth[6] (World Bank, 2020) in the model. GHG emissions are associated with a country's level of development. As developing countries grow their economies, that is, industrialize, GDP per capita and GHG emissions are expected to increase. GHG are negative externalities if the production process. Thus, increases in GDP per capita growth over time indicate the extent to which the countries in the sample have industrialized and indicate the extent to which GHG emissions is associated to this industrialization process. Moreover, I include measures that capture the impact of different economic sectors to a country's economy. Different economic sectors impact GHG emissions differently and the inclusion of these variables isolates the participation of each sector in a country's total GHG emission. The agriculture variable portrays the percentage participation of the agricultural, forest, and fishing sectors in the economy, in a year[7] (World Bank, 2020). The industry variable comprises value added in mining, manufacturing, construction, electricity, water, and gas, in a year[8] (World Bank, 2020).

The panel dataset contains observations of the twenty countries in the sample during the period from 1990 to 2015. Panel data often shows a specific type of statistical errors, non-spherical errors, which result from of contemporaneous correlation across the units. This statistical error refers to the difference between a value described in the dataset and the actual value. To deal with this type of error, I estimate an ordinary least squares with panel-corrected standard errors model, which is robust to non-spherical errors (Beck & Katz, 1995, 2011).

Considering that the dependent variable may present time dependency issues, I test its stationarity. Stationarity is a characteristic of time series that do not show an autocorrelation structure or, in other words, periodic fluctuations over time. Results of unit-root tests of the dependent variable, GHG emissions, indicate that the variable is stationary, or not autocorrelated, and thus, when estimating the effect of EPS on GHG emissions, I do not adjust the models for autocorrelation. Yet, I adjust them for panel-level heteroskedasticity. Heteroskedasticity refers to the unequal variance of GHG emissions and as GHG emissions vary substantially across countries and years, I adjust the model, Moreover, as unit-root tests of the independent variable, EPS, indicate that the variable has a unit root, I include a 1-year lag in the analysis.

STATISTICAL FINDINGS

To investigate the extent to which EPS affects GHG emissions, I run three ordinary least squares models with panel-corrected standard (OLS PCSE). Models 1, 2 and 3 examine the impact of environmental policies on GHG emissions. As evidence suggests that different combinations of regulations produce better environmental outcomes (Caputo, 2014; Hagner, 2000; Harrison et al., 2015), I do not focus on the distinct effects of either market or nonmarket regulations, but on the EPS index. The dependent variable is total GHG emissions. As this is a stationary variable, it is not lagged. However, as unit-root tests of the EPS index, the independent variable, indicate that environmental policies have unit-roots, it is lagged. Models 1, 2 and 3 include 1-, 3-, and 5-year lags of the index.

Results of the three models are similar. As expected, EPS is inversely proportional to GHG emissions. Increases in stringency are associated with decreases of GHG emissions. In model 1, an increase of 1 unit in the EPS index is associated with a decrease of 0.164 standard deviation of total GHG emissions in a year, all else constant. In model 2, the increase is accompanied by decrease of 0.169 standard deviation of total GHG emissions in a year, all else constant, and model 3, which includes the EPS index with a 5-year lag, shows a slightly more pronounced decrease in GHG emissions levels. An increase of 1 unit in the EPS index is correlated with a decrease of 0.195 standard deviation of total GHG emissions in a year, all else constant. Findings in models 1, 2, and 3 are in accordance with expectations regarding the outcomes of environmental policies. They demonstrate that EPS is negatively associated with GHG emissions and that policy outcomes take time to produce optimal results.

Moreover, increases in the percentage participation of the agricultural sector in a country's GDP are associated with a decrease in GHG emissions. On the other hand, industrial value added in mining, manufacturing, construction, electricity, water, and gas is associated with increases in GHG emissions, as expected. Similarly, the annual percentage of GDP per capita growth is also associated with increases in GHG emissions. Contrary to expectations, investments in research and development are associated with increases in GHG emissions. I anticipated that these investments would promote green technology development and innovation that could cancel out the production of negative externalities. It is possible, however, that they will still promote environmental conservation with time. The expected impacts of subsidies to R&D on GHG emissions could be lagged; it remains to be seen. Nevertheless, this relationship presents a puzzle.

In summary, evidence demonstrated by results of the OLS PCSE models supports the hypothesis; more stringent environmental policies are associated with improvements in environmental quality. The magnitude of such improvements, however, is small.

	Model 1 b/se	Model 2 b/se	Model 3 b/se
Dependent Variable: *Total GHG emissions (Standardized)*			
EPI Index, 1 year lagged	-0.164**		
	(0.6)		
EPI Index, 3 years lagged		-0.169*	
		(0.07)	
EPI Index, 5 years lagged			-0.195*
			(0.08)
Agricuture (Standardized)	-0.239**	-0.238**	-0.240**
	(0.09)	(0.09)	(0.09)
Industry (Standardized)	0.650***	0.655***	0.651***
	(0.08)	(0.08)	(0.08)
Total GDP (Standardized)	0.205***	0.199***	0.208***
	(0.05)	(0.06)	(0.06)
Subsidies R&D	0.401***	0.393***	0.391***
	(0.07)	(0.07)	(0.07)
Constant	-0.303***	-0.311***	-0.306**
	(0.09)	(0.1)	(0.1)
R2	0.41	0.41	0.41
Number of obs.	288	288	285

Figure 11.2 Statistical Models. (OLS PCSE). * p<0.05, ** p<0.01, *** p<0.001. *Source*: This figure has been generated by the author.

CONCLUDING THOUGHTS

Have environmental policies promoted environmental quality? The analysis of the EPS index in twenty of the world's largest countries, responsible for about 80 percent of the global GDP (World Bank, 2020) and 80 percent of total GHG emissions (OECD, 2018), suggests that stringent policies are associated with a small reduction in GHG emissions. Stringent policies seem to have incrementally promoted environmental quality.

In the end, the results of this analysis indicate that environmental policies have been marginally effective in promoting environmental quality, and thus, suggest that governments are on the track to policy effectiveness. Nevertheless, results do not indicate that governments will alter the course of climate change with the examined environmental policies. A lot of work is still necessary and the feedback from analyses like this must be incorporated into policy-making if we, as a global society, are to change the course of the climate crisis. The impact of environmental policies must be unraveled. Even though studies have suggested that different combinations of environmental regulations are best at mitigating environmental externalities (Caputo, 2014; Hagner, 2000; Harrison et al., 2015), the effect of each type of regulation on GHG emissions must be better understood. By investigating how specific types of regulations impact GHG emissions, researchers may figure out the most effective way to curb them.

Nevertheless, policy outcomes take time to be fully observed. A time span of twenty-five years is still considered a short period for policy outcome analyses. Consistent with this idea, findings indicate that slightly more pronounced decreases in GHG emissions levels have occurred five years after their implementation. Therefore, it remains to be seen whether these policies will be associated with future decreases in GHG emissions. Up until now, decreases are small and not sufficient to change the course of the climate crisis.

Some questions remain unanswered and could be investigated in future research. For instance, considering the level of trade liberalization in the world, do environmental policies implemented in one country impact GHG emissions in another? If the relationship between EPS in a country and GHG emissions in another is validated, EPS could be associated with the promotion of more environmental quality.

Moreover, analyses of the economic consequences of environmental policies must also inform environmental policy decisions. Yet, economic, trade, and industrial policies must address the fight to mitigate the negative impacts of climate challenge. Innovative economic policies that promote a reduced environmental impact must be sponsored.

Different avenues of GHG emission reduction should be considered. Apart from government actions, could multinational corporations and individuals play a role in the fight to alter climate change? If the climate change challenge is taken seriously, governments, businesspeople, the scientific community, and the public must act to find innovative and sound solutions.

NOTES

1. The example refers specifically to American society since consumption levels in the United States are among the highest in the world. The United States has about 4.7 percent of the world's population and emits close to 15 percent of total GHG emissions (OECD, 2020a). Nevertheless, as the American Way of Life is a model for development and growth, several societies have mirrored this pattern and got closer to American consumption levels, especially since the 1990s (For more information, see Soto, 2000).
2. For some critical perspectives, see the works of Emerson (1975) and Thoreau (1997).
3. Feed-in-tariffs (FITs) support scaling up renewable electricity capacity. FITs offer long-term contracts that guarantee a price to be paid to source of electricity per kWh fed into the electricity grid (OECD, 2017).
4. Deposit & Refund Scheme (DRS) is the surcharge on the price of potentially polluting products. If pollution is avoided, a refund of the surcharge is granted (OECD, 2017).
5. The dependent variable, total GHG emissions, is standardized. A standardized variable is rescaled to have a mean of zero and a standard deviation of one. The standardization process puts different variables on the same scale and produces more reliable results.
6. Total GDP is standardized (see note 5).
7. Agriculture is standardized (see note 5).
8. Industry is standardized (see note 5).

REFERENCES

Arndt, C., Chinowsky, P., Robinson, S., Strzepek, K., Tarp, F., & Thurlow, J. (2012). Economic Development under Climate Change. *Review of Development Economics*, *16*(3), 369–377. https://doi.org/10.1111/j.1467-9361.2012.00668.x

Barnett, M., & Finnemore, M. (2004). *Rules for the World: International Organizations in Global Politics*. Cornell University Press.

Capstick, S. (2013). Public Understanding of Climate Change as a Social Dilemma. *Sustainability*, *5*(8), 3484–3501. https://doi.org/10.3390/su5083484

Collins, C. (2020, February 6). Six Texas Oil Refineries Are Among the Nation's Worst Benzene Polluters, Data Shows. *The Texas Observer*. https://www.texasobserver.org/benzene-oil-refineries-texas-coast/

Dąbrowska, J., Pawęska, K., Wrocław University of Environmental and Life Sciences, Dąbek, P. B., Wrocław University of Environmental and Life Sciences,

Stodolak, R., & Wrocław University of Environmental and Life Sciences. (2017). The Implications of Economic Development, Climate Change and European Water Policy on Surface Water Quality Threats. *Acta Scientiarum Polonorum Formatio Circumiectus*, *3*, 111–123. https://doi.org/10.15576/ASP.FC/2017.16.3.111

Denny Ellerman, A., & Montero, J.-P. (2007). The Efficiency and Robustness of Allowance Banking in the U.S. Acid Rain Program. *The Energy Journal*, *28*(4). https://doi.org/10.5547/ISSN0195-6574-EJ-Vol28-No4-3

Elkins, Z., Guzman, A. T., & Simmons, B. A. (2006). Competing for Capital: The Diffusion of Bilateral Investment Treaties, 1960–2000. *International Organization*, *60*(04). https://doi.org/10.1017/S0020818306060279

Emerson, R. W. (1975). *Essays: First and second series*. Houghton, Mifflin. http://books.google.com/books?id=DpRaAAAAMAAJ

Gonzalez-Perez, M.-A., & Leonard, L. (Eds.). (2017). *Climate Change and the 2030 Corporate Agenda for Sustainable Development* (1st edition). Emerald.

Gourevitch, P. (1978). The Second Image Reversed: The International Sources of Domestic Politics. *International Organization*, *32*(4), 881–912. https://doi.org/10.1017/S002081830003201X

Jacques, P. J., & Knox, C. C. (2016). Hurricanes and Hegemony: A Qualitative Analysis of Micro-Level Climate Change Denial Discourses. *Environmental Politics*, *25*(5), 831–852. https://doi.org/10.1080/09644016.2016.1189233

Johnson, B. B., & Finkel, A. M. (2016). Public Perceptions of Regulatory Costs, Their Uncertainty and Interindividual Distribution: Public Perceptions of Regulatory Costs. *Risk Analysis*, *36*(6), 1148–1170. https://doi.org/10.1111/risa.12532

Joskow, P. L., & Schmalensee, R. (1998). The Political Economy of Market-Based Environmental Policy: The U.S. Acid Rain Program. *The Journal of Law and Economics*, *41*(1), 37–84. https://doi.org/10.1086/467384

Kingdom, J. (1993). Politicians, Self-Interest and Ideas. In G. E. Marcus and R. L. Hanson (Eds.). *Reconsidering the democratic public* (pp. 73–89). Pennsylvania State University.

Lewandowsky, S., Oreskes, N., Risbey, J. S., Newell, B. R., & Smithson, M. (2015). Seepage: Climate Change Denial and Its Effect on the Scientific Community. *Global Environmental Change*, *33*, 1–13. https://doi.org/10.1016/j.gloenvcha.2015.02.013

Mayhew, D. R. (2004). *Congress: The electoral connection* (2nd edition). Yale University Press.

McGillivray, F. (2004). *Privileging industry: The comparative politics of trade and industrial policy*. Princeton University Press.

Mullin, R. P. (2002). What Can Be Learned from DuPont and the Freon Ban: A Case Study. *Journal of Business Ethics*, *40*(3), 207–218. https://doi.org/10.1023/A:1020511815499

OECD. (2017). *Environmental Policy Stringency Index* (Edition 2017) [Data set]. OECD. https://doi.org/10.1787/b4f0fdcc-en

OECD. (2018). *Air and GHG Emissions* [Data set]. OECD. https://doi.org/10.1787/93d10cf7-en

OECD. (2020a). *Environment Statistics* [Data set]. OECD. https://www.oecd-ilibrary.org/environment/data/patents-in-environment-related-technologies/technology-indicators_e478bcd5-en

OECD. (2020b). *Green Growth Indicators* [Data set]. OECD Environment Statistics (database). OECD. https://doi.org/10.1787/data-00665-en

Porter, M. E., & Linde, C. van der. (1995). Toward a New Conception of the Environment-Competitiveness Relationship. *Journal of Economic Perspectives, 9*(4), 97–118. https://doi.org/10.1257/jep.9.4.97

Putnam, R. D. (1988). Diplomacy and Domestic Politics: The Logic of Two-Level Games. *International Organization, 42*(3), 427–460. https://doi.org/10.1017/S0020818300027697

Schmalensee, R., & Stavins, R. N. (2013). The SO2 Allowance Trading System: The Ironic History of a Grand Policy Experiment. *Journal of Economic Perspectives, 27*(1), 103–122. https://doi.org/10.1257/jep.27.1.103

Shukla, P. R., Skea, J., Buendia, E. C., Masson-Delmotte, V., Portner, O., Roberts, D. C., & Zhai, P. (Eds.). (2019). *Climate Change and Land.* Intergovernmental Panel on Climate Change. https://www.ipcc.ch/srccl/

Silver, L. (2019). *Smartphone Ownership Is Growing Rapidly Around the World, but Not Always Equally.* Pew Research Center. https://www.pewresearch.org/global/2019/02/05/smartphone-ownership-is-growing-rapidly-around-the-world-but-not-always-equally/

Simmons, B. A., & Elkins, Z. (2004). The Globalization of Liberalization: Policy Diffusion in the International Political Economy. *American Political Science Review, 98*(1), 171–189. https://doi.org/10.1017/S0003055404001078

Smith, B. (1998). Ethics of Du Pont's CFC Strategy 1975-1995. *Journal of Business Ethics, 17*(5), 557–568.

Soto, M. (2000). *Capital Flows and Growth in Developing Countries: Recent Empirical Evidence* (Working Paper No. 160). OECD Development Center.

Thoreau, H. D. (1997). *Walden.* Oxford University Press.

United Nations. (1987). *Montreal Protocol on Substances that Deplete the Ozone Layer.* United Nations. https://treaties.un.org/doc/publication/unts/volume%201522/volume-1522-i-26369-english.pdf

World Bank. (2020). *World Development Indicator (WDI) Databank.* World Bank. https://data.worldbank.org/

Zahariadis, N. (2007). The Multiple Streams Framework: Structure, Limitations, Prospects. In P. A. Sabatier (Ed.), *Theories of the Policy Process* (2nd edition). Westview Press.

Chapter 12

The Role of Art in the Conservation of American Landscapes

Joe R. McBride

The European colonists who came to the eastern shores of America were confronted by landscapes of seemingly endless forests. The pilgrims in New England found deciduous forests supporting oak, chestnut, beech, and maple trees. The English settler in the mid-Atlantic was confronted by loblolly pine and oak forests, and the Spanish colonists in Florida encountered forests of longleaf pine (Bonnicksen, 2000). They saw in these forests unlimited resources of timber and firewood as well as dangers associated with wild animals and Native Americans. Forests were cleared for village sites and farming, often with wide borders for defensive purposes. Trees were felled for building material and also utilized for ship building and the production of naval stores. The perception of danger associated with the forest went beyond the physical threats to include other psychological threats common to Europeans at that time. In the early part of the fourteenth century, Dante Alighieri (2019) composed his famous narrative poem *The Divine Comedy*. In it, he describes his journey through hell, purgatory, and paradise. His famous introductory line is as follows: "In the middle of the journey of our life I found myself with a dark forest where the straightforward pathway had been lost" (Alighieri, 2019). It reflects the dangers of the forest as perceived by Europeans during the Middle Ages. Dante used the fear of physically getting lost in a forest to the dilemma of morally becoming lost. Three centuries later, John Bunyon's *The Pilgrim's Progress* (1975) used the dark woods of the "Valley of the Shadow of Death" as metaphors for hell, darkness, and terror. His "Slough of Despond," a miry swamp, served as a hazard on his journey where a man falling into it sank under the weigh of his sins and his sense of guilt, as a traveler might sink into an actual swamp. *Pilgrim's Progress*, widely read by American Puritan colonists in the late seventeenth century, reinforced a moral fear of the forest.

With the passage of time, the early colonist's perceived views of the forest's endless resources and the eminent dangers of the forests would change. These changes were due to the overexploitation of the forests and to the influence of American artist on the perception of the forest. Conservation of the America's forests and scenic landscapes was propelled by works of art starting in the early nineteenth century. The influence of art on the conservation movement continued in the twentieth century as it does today. The objective of this chapter is to explore the role of art in the conservation of America's forests and wild landscapes.

ART AND CONSERVATION IN COLONIAL AMERICA

Painting during the colonial period in America was almost exclusively focused on portraits of well-to-do individuals and families. Landscape paintings and paintings of forests were not commissioned or sought after by those wealthy enough to afford them. There were, however, portraits composed on balconies or in rooms with windows behind the human subjects to show their wealth in terms of the land they owned or simply a glimpse of the landscape. Occasionally, trees and forests were included in the views through the windows. These colonial portraits did little to influence the attitude of Americans about the scenic or intrinsic values forests and wild landscapes. They did, however, reinforce the economic values of the land.

Efforts began as early as 1631 to conserve forest resources in America when the Massachusetts Bay Colony forbade the burning of any ground prior to March 1 in order to protect tree seedlings. In 1681 William Penn, in the Pennsylvania Colony, required that for every 5 acres of forest cleared 1 acre should be kept for trees (Dana & Fairfax, 1980). The "Broad Arrow" policy was created by the British government in 1691 to forbid the cutting of all trees over 24 inches in diameter on land not granted to a private person (Malone, 1965). Trees of this size were marked with slashes resembling a broad arrow and reserved for the exclusive use of the Crown for ship building. In 1729, the "Broad Arrow" policy was reenacted with somewhat stricter provisions as to what constituted private land. The strict enforcement of the "Broad Arrow" policy has been identified as a factor contributing to the American Revolution (Dana & Fairfax, 1980). Additional efforts by the British crown to protect forest producing naval stores needed for shipbuilding were made by imposing penalties for cutting of protected trees; however, no efforts were made to protect land for its scenic or intrinsic value.

A new generation of American painters emerged with an interest in painting landscapes in the first half of the nineteenth century. They had an important influence on the attitude of Americans toward forests and nature.

Among these artists was Thomas Cole, an English-born American painter who is regarded as the founder of the "Hudson River School" (Truettner, 1994). The term "Hudson River School" has been applied to a group of artists whose work focused on the landscapes in and around the Hudson River, although they did not work together and were in no sense a "school" of painters. Prominent among these artists, in addition to Cole, were Frederick Edwin Church, John Frederick Kensett, Sanford Edwin Gifford, and Asher Brown Durand. Thomas Cole's painting *A View of the Two Lakes and Mountain House, Catskill Mountains, Morning*, 1844 (figure 12.1) is typical of his capturing a forest landscape in the Hudson Valley region. The painting invites us to walk with the man in the foreground along the ridge through the forest to the Mountain House. The forest is not menacing and does not evoke the dangers proposed by Dante Alighieri and John Bunyon. The painting was done at a time when much of the Hudson River Valley and New England had been logged, often by clearcutting (Raup, 1966). However, Cole and other members of the Hudson Valley School focused their attention on landscapes that were still wild.

The non-threatening images of the American forests painted by the Hudson Valley School were reinforced by the writings of the American poets and transcendentalists in the first half of the nineteenth century. William Cullen Bryant (1794–1878) extolled the sacredness of the forest in his poem *A*

Figure 12.1 Thomas Cole (American, born England, 1801–1848). A View of the Two Lakes and Mountain House, Catskill Mountains, Morning, 1844. *Source*: Oil on canvas, 35 13/16 × 53 7/8 in. (91 × 136.9 cm). Brooklyn Museum, Dick S. Ramsay Fund, 52.16.

Forest Hymn (Bryant, 2013). The first line of that poem written in 1824 is as follows: "The groves were God's first temples." Ralph Waldo Emerson (1803–1882) and Henry David Thoreau (1817-1862) promoted the forest as places where one could learn in their essays and journals (Emerson, 2003; Thoreau, 2010). In 1836, Emerson wrote in his essays on nature: "In the woods, we return to reason and faith" (Emerson, 2003). Thoreau famously wrote in 1854: "I went to the woods because I wished to live deliberately, to front only the essential facts of life, and see if I could not learn what it had to teach, and not, when I came to die, discover that I had not lived" (Thoreau, 2010).

Asher Brown Durand depicted the painter Thomas Cole and the poet William Cullen Bryant in his painting *Kindred Spirits*. The painting illustrates a peaceful forest that has inspired both the painter and the poet. Like other paintings, poems, and essays it contributed to a different, more positive view of forests and wild landscapes. This shift in the way Americans thought of their forests did not have an immediate impact on forest or landscape conservation in America, but it was fundamental in creating a mindset that would support a conservation movement in the second half of the nineteenth century.

In the 1820s and 1830s tourism became fashionable in America, especially in areas near major metropolitan areas. Roads, trails, and accommodations were developed for tourists. Much of the tourist's interest was focused around viewing wild landscapes (Johnson, 1990). A synergy between the tourists and Hudson Valley School painters developed during the first half of the nineteenth century. Well-to-do tourists wanted paintings of the wild landscapes they had visited, and painters wanted to market their paintings. The exhibiting of paintings in New York and other cities spurred more interest in tourism and increased tourism spurred the demand for painting of the wild landscapes.

During the first half of the nineteenth century, private land was purchased and federal land set aside to protect forests that were fit to produce timber and naval stores for ship construction. Small areas of land were purchased along the Georgia coast following the appropriation of $200,000 by Congress in 1799. In 1817, another congressional act directed the reservation of public lands bearing live-oak or cedar timber suitable for the Navy, as might be selected by the president (Fernow, 1951). Under these and subsequent acts in 1828 and 1831 some 244,000 acres of forest land were reserved in Alabama, Florida, Louisiana, and Mississippi. As with the "Broad Arrow Act" of the British crown, this legislation was aimed at protecting trees for future naval construction, not for the preservation of wild landscapes for their scenic, intrinsic, or recreational values.

PAINTING AND PHOTOGRAPHING THE AMERICAN WEST

Exploration of the west led to the discovery of extensive forest and wild landscapes of amazing beauty. Government and private exploration parties often included artists who recorded the landscape, plants, animals, and indigenous people in the age before the camera. For example, the American artist Rembrandt Peale served as the official artist on the U. S. government's exploration to find the headwaters of the Arkansas and Red rivers led by Stephen Long in 1820 (Beidleman, 1986). The Swiss artist, Karl Bodmer, accompanied Prince Maximillian of Wied-Neuwied on his exploration of portions of the upper Missouri River in 1834-34 (Ewers, 1984). George Catlin independently mounted his own expedition into the west in order to create a number of paintings to support his profession as an artist (Catlin, 1863).

Yosemite Valley was discovered by white men in 1851. News of the spectacular valley spread slowly in California. In 1855 the artist Thomas Ayres accompanied James Mason Hutchings on a trip into Yosemite Valley. Hutchings sponsored the trip in order to gather information and illustrations for a magazine he planned to publish. Ayres's drawings were the first images to be recorded of the Yosemite landscape (figure 12.2). They appeared in Hutchings magazine and were distributed nationally. An exhibit of Ayres's drawings was held in New York City. This exhibit along with the writings

Figure 12.2 Thomas Ayres. 1855. *Yosemite Valley*. *Source*: Yosemite Museum Research Library. (Courtesy of the Yosemite National Park Archives, Museum, and Library.)

of Hutchings and Ayres inspired considerable interest in the newly discovered valley as a destination for tourists. In 1861 the outdoor photographer Carelton Watkins traveled to Yosemite to take the first photographs of valley (figure 12.3). His photographs were exhibited at a gallery in New York in 1862. *The New York Times* reported that "The views of lofty mountains, of gigantic trees, of waterfalls . . . are indescribably unique and beautiful" (Naef & Wood, 1975). Watkins's photography contributed to efforts by California boosters to promote the state by setting aside land in Yosemite Valley and the nearby Mariposa Big Tree Grove. California Senator John Conness introduced a bill in Congress to protect the area. In 1864 President Lincoln signed the Yosemite Valley Grant Act, Senate Bill 203, to remove Yosemite Valley and the nearby Mariposa Big Tree Grove from the federal land. This legislation transferred the land to the state of California "upon the express conditions that the premises shall be held for public use, resort, and recreation" (Heckscher, 1996). It is believed that Conness, who owned a collection of Watkins's photographs and was a friend of Lincoln, showed the photos to the president to convince him to sign the bill protecting Yosemite (Jarvis, 2016).

Albert Bierstadt, a German-born American artist, who was initially associated with the Hudson River School travelled west in 1863 to visit Yosemite

Figure 12.3 Carlton Watkins. 1861. *Yosemite Valley*. *Source*: Library of Congress. (Used courtesy of the Library of Congress.)

Valley with his friend, the writer, Fitz Hugh Ludlow. Their trip was believed to have been inspired by Watkin's photographs of Yosemite. They spent several weeks in Yosemite Valley where Bierstadt made landscape sketches that he later expanded into large-scale paintings in his studio in New York. These paintings were exhibited in theater-like galleries in Manhattan where hundreds of New Yorker were able to view images of the western landscapes for the first time (Anderson & Ferber, 1990). Among those paintings exhibited was Beirstadt's *Valley of the Yosemite* (figure 12.4). A broader American audience was made aware of western scenery by an article published by Ludlow in the June 1864 issue of Atlantic Monthly. In it he proclaimed that Yosemite Valley surpassed the Alps in water falls and the Himalayas in precipices (Ludlow, 1864). Neither Bierstadt's paintings nor Ludlow's article had a direct influence on the setting aside of Yosemite Valley and the Mariposa Big Tree Grove, as they appeared after Lincoln had signed the Yosemite Valley Grant Act. They did, however, inform Americans about the scenic wonders of the west and contributed to a growing interest in conserving areas of outstanding landscapes.

In 1871, the Hayden Geological Survey explored the Yellowstone area (Merrill, 1999). The photographer William Henry Jackson and the painter Thomas Moran accompanied the expedition. Jackson's photographs and Moran's paintings inspired Congress to withdraw the area from public auction and to set it aside as America's first National Park. Moran's painting *The Grands Canyon of the Yellowstone* (figure 12.5) was exhibited under the rotunda of the

Figure 12.4 ***Albert Bierstadt, Looking up Yosemite Valley*. 1863.** *Source*: Haggin Museum, Stockton, CA. (Used with permission of the Haggin Museum.)

Figure 12.5 Thomas Moran. *The Grand Canyon of the Yellowstone*. 1872. *Source*: Smithsonian American Art Museum. (Used courtesy of the Smithsonian American Art Museum.)

capitol building during the congressional debate by supporters of the act to set aside the land as a park (Greenwald, 2015). Members of Congress had to pass by the painting in order to reach Senate and House of Representatives chambers. President Grant signed the Act of Dedication that created Yellowstone National Park in 1872. Moran subsequently painted various locations in the west that were later to become National Park or National Monuments (Kinsey, 1992). These paintings promoted the preservation of these landscapes as well as tourism. Tourism played an important part in the development of the political pressure necessary to protect and preserve these areas.

John Wesley Powell explored the Grand Canyon in his initial journey in 1869 (Powell, 1875). No photographs were made during that expedition. He returned to the Grand Canyon for a second expedition in 1871 accompanied by photographers who took over 100 photographs. Powell used the photographs to illustrate his lectures tours. These tours made Powell a national figure and brought attention to the Grand Canyon. Efforts by U.S. Senator Benjamin Harrison in 1877 to establish the Grand Canyon as a National Park failed in Congress, but Harrison set the area aside as a Forest Preserve in 1893 when he became president (Uenuma, 2019). Theodore Roosevelt redesignated the area as a National Monument in 1908 and it became a National Park in 1919. Photographs taken during the second Powell expedition can be credited along with paintings by Thomas Moran in the eventual establishment of Grand Canyon National Park.

A host of other American artists painted and photographed wild American landscapes in the latter half of the nineteenth century. The work of artists such as George Caleb Bingham, Charles Deas, Thomas Hill, Alfred Jacob Miller, Charles Marion Russell, Frederick Remington, and photographers like William Bell, Edward Sherriff Curtis, William Henry Jackson, Timothy O'Sullivan, and Andrew Joseph Russell contributed to an appreciation by the public of the scenic beauty and grandeur of western landscapes. The work of artists and photographers was fundamental to the development of political support for the protection of the wild landscapes of the west.

The early decades of the twentieth century saw the setting aside scenic areas of the American landscape through the establishment of national parks and monuments, wilderness areas, state parks, and other protective measures. These measures owed much to the influence of the nineteenth-century artists, photographers, and authors on changing the attitudes of the American public about wildlands. Support for the protection of American landscapes also came from a number of authors like George Perkins Marsh, John Muir, and John Burroughs as well as political leaders like Theodore Roosevelt, Benjamin Harrison, and Woodrow Wilson (Nash, 1982). Passage of the Antiquities Act of 1906 allowed the president to proclaim National Monuments. Theodore Roosevelt used this act to set aside a number of National Monuments in the west. In 1916 President Wilson signed an act creating the National Park Service and in 1933 an Executive Order transferred fifty-six national monuments and military sites from the Forest Service and the War Department to the National Park Service. Today some 84.6 million acres of land is under management by the National Park Service. Aldo Leopold, a Forest Service Supervisor, convinced the Forest Service to set aside as wilderness area of 500,000 acres of the Gilia National Forest in New Mexico. This initiated further protection of wilderness landscapes in National Forests. In 1964 the Wilderness Act was signed by President Johnson which has resulted in the protection of 9.1 million acres of wilderness.

Visual representation was an important factor in the setting aside of land for national parks, national monuments, and wilderness areas. It gained the support of the public and politicians for setting land aside for its recreational, scenic, and intrinsic values. Dependence on visualization did not, however, end in the first half of the twentieth century. Photography was to play an important role in the further preservation of Americans landscapes in the latter half of the century. Efforts by the Save the Redwoods League and the Sierra Club to establish a Redwood National Park in California were greatly aided by the photography of Ansel Adams and Philip Hyde (Leydet, 1969). Adams's photograph, *Redwoods, Bull Creek Flat, Northern California*, presented a stunning view of redwood trees. Although the interior of the forest could not be seen in the photograph, there is no doubt that the American public saw anything threatening in the photograph (figure 12.6). Philip Hyde's photographs, used to illustrate the book *The Last Redwoods* (Hyde & Leydet, 1963), illustrated the last stands of the redwood forest and the impact of commercial logging. Photographs by these two photographers were used in an intense effort to lobby Congress to pass a bill to establish the Redwood National Park (Schrepfer, 1983). The bill was passed and signed by President Lyndon Johnson in 1968.

Current controversaries over the size of certain National Monuments or the ability of petroleum companies to drill for oil on previously protected federal land, such as the Arctic National Wildlife Refuge, are being fought with photographic images of the landscapes in question (Banjerjee, 2003). These controversaries are not over, but it can be anticipated that public support for the continued protection of these lands will depend very much on the artistic representation of these landscapes.

Figure 12.6 Ansel Adams. *Redwoods, Bull Creek Flat, Northern California*. 1960. *Source*: Chrysler Museum of Art. Norfolk, VA. (Used with permission from the Chrysler Museum of Art.)

ACKNOWLEDGEMENTS

I would like to express my appreciation to Professor Margaretta Lovell for the knowledge she conveyed to me while we jointly taught a course on the American forest at the University of California, Berkeley. From her lectures and our conversations, I gained a new understanding of the history of American art and its significance to the development of forest conservation in the United States. This book chapter is a direct result of her tutoring

REFERENCES

Alighieri, D. (2019). *The divine comedy* (A. Mandelbaum, Trans.). Everyman's Library.

Anderson, N. K. & Ferber, L. S. (1990). *Albert Bierstadt, art and enterprise.* Brooklyn Museum.

Banerjee, S. (2003). *Arctic National Wildlife Refuge: Season of Life and Land.* Braided River.

Beidleman, R. G. (1986). The 1920 Long Expedition. *American Zoology, 26*, 307–313.

Bonnicksen, T. M. (2000). *America's ancient forests.* Wiley and Sons.

Bryant, W. C. (2013). *A forest hymn.* HardPress Publishing.

Bunyan, J. (1975). *The Pilgrim's progress* (R. Sharrock and J. B. Wharey Eds.). Oxford University Press.

Catlin, G. (1861). *Life among the Indians.* Gall and Inglis.

Dana, S. T., & Fairfax, S. (1980). *Forest and range policy: Its development in the United States.* McGraw-Hill.

Emerson, R. W. (2003). *Nature and selected essays.* Penguin Books.

Ewers, J. C. (1984). *Views of vanishing frontier.* Joslyn Art Museum.

Fernow, B. E. (1951). *A story of North American forestry.* Princeton, NJ: Princeton University Press.

Greenwald, D. S. (2015). The big picture: Thomas Moran's *The Grand Canyon of the Yellowstone* and the development of the American West. *Winterthur Portfolio: A Journal of American Material Culture, 49*(4), 175–210.

Heckscher, J. J. (1996). *The evolution of the conservation movement, 1850–1920.* The Library of Congress.

Hyde, P., & Leydet, F. (1963). *The last redwoods. Photographs and story of a vanishing scenic resource.* Sierra Club.

Jarvis, B. (2016, June). How an obscure photographer saved Yosemite. *Smithsonian Magazine.* https://www.smithsonianmag.com/history/carleton-watkins-yosemite-photographer-national-parks-180959065/.

Johnson, K. (1990). Origins of tourism in the Catskill Mountains. *Journal of Cultural Geography, 11*(1), 5–16.

Kinsey, J. L. (1992). *Thomas Moran and the surveying of the American West.* Smithsonian Institution Press.

Leydet, F. (1969). *The last redwoods, and Parklands of Redwood Creek*. Sierra Club/Ballantine Books.

Ludlow, F. H. (1864, June). Seven weeks in the great Yosemite. *Atlantic Monthly, 13*(80), 739–54.

Merrill, M. D. (Ed.). (1999). *Yellowstone and the great West: Journals, letters and images from the 1871 Hayden expedition*. University of Nebraska Press.

Naef, W. J., & Wood, J. N. (1975). *Era of exploration: The rise of landscape photography in the American West, 1860–1885*. New York Graphic Society

Nash, R. (1982). *Wilderness and the American mind*. Yale University Press.

Malone, J. J. (1965). *Pine trees and politics*. University of Washington Press.

Powell, J. W. (1875). *The exploration of the Colorado River and its canyons*. Dover Press.

Raup, H. M. (1966, April). The view from John Sanderson's farm: A perspective on the use of the land. *Forest and Conservation History, 10*(1), 2–11.

Schrepfer, S. R. (1983). *The fight to save the redwoods: A history of environmental reform, 1917–1978*. University of Wisconsin Press.

Thoreau, H. D. (2010). *Walden*. Shambhala Publications.

Truettner, W. H. (1994). *Thomas Cole: Landscape into history*. Yale University Press.

Uenuma, F. (2019, February 26). The decades-long political fight to save the Grand Canyon. *Smithsonian Magazine*. https://www.smithsonianmag.com/history/decades-long-political-fight-save-grand-canyon-180971567/.

Index

Page references for figures and tables are italicized

About the Contributors

Emilia Barreto Carvalho is an assistant professor of political science at Lone Star College, Tomball in Texas. She is interested in international political economy with a focus on the environment. Her research examines the international diffusion of environmental policies, the factors impacting policy implementation by states, and policy outcomes. Emilia is currently working to further incorporate the United Nations' Sustainable Development Goals on the campus where she teaches. In her free time, she volunteers with a recycling project at her children's school and has recently started a small victory garden at home.

Alan C. Clune completed his BS at Worcester Polytechnic Institute, his MS in philosophy at Rensselaer Polytechnic Institute, and his PhD in philosophy at the State University of New York at Buffalo. He is a philosophy professor at Sam Houston State University. His current research interests are primarily in applied ethics with a focus on animals and ethics.

John A. Duerk is a tenure-track history and political science professor at Lake Tahoe Community College in South Lake Tahoe, California. Previously, he served as the chair of the Government Department and coordinator of the Center for Civic Engagement at Lone Star College, CyFair in the Houston suburbs. His career in education began more than twenty years ago when he taught at the secondary level in Illinois. John's teaching philosophy is rooted in Robert Putnam's work on social capital, and his teaching interests include governing institutions, political participation, and a variety of policy issues such as privacy and economic inequality. Furthermore, he is a strong advocate of viewpoint diversity to develop a more thorough understanding of public problems. His article, "Elijah P. Lovejoy: Anti-Catholic Abolitionist," appeared in the peer reviewed *Journal of the Illinois State Historical Society*. John holds a PhD in political science and an MA in history, both from Northern Illinois University.

Dr. Jennifer Epley Sanders is an associate professor of political science at Texas A&M University, Corpus Christi. She holds an MA and PhD in political science from the University of Michigan-Ann Arbor, as well as a certificate in Southeast Asian studies and a certificate in survey methodology. Her BA in political science is from Vassar College in New York. Her areas of research specialization include comparative government and politics; Southeast Asia; identity and politics with an emphasis on religion and gender; public opinion and political behavior; and research methodology. She has lived and worked abroad, particularly in Indonesia and Australia, for several years. Dr. Epley Sanders also conducts research and publishes articles on the Scholarship of Teaching and Learning. She teaches core curriculum courses such as Introduction to U.S. Government and Politics and political science major classes like Comparative Politics, International Relations, and Religion and Politics.

Shelby Hockaday received her BS in earth sciences (geography) from Oregon State University where she studied conflict-negotiation for water resources. She is a social scientist with interests in legal geography, western water law, and adaptive management of water resources. Having previously worked in project management for an international development contractor, Shelby is currently a master's student in the Department of Geography at the University of Nevada, Reno, researching western water law and adaptive responses to water issues such as resource scarcity, changing social values, and climate variability.

Dr. Elizabeth Koebele is an assistant professor of political science at the University of Nevada, Reno. She is also affiliated faculty in the Graduate Program for Hydrologic Sciences. As an interdisciplinary social scientist, Elizabeth researches environmental policymaking processes, with a focus on water governance in the western United States. She is particularly interested in how collaborative approaches to environmental governance shape both the policy process and socio-environmental outcomes. Elizabeth is currently conducting research on adaptations to climate change in the water resources sector, sustainable urban water management, and disaster resilience policy.

Dr. Scott A. Lukas is faculty chair of Teaching and Learning at Lake Tahoe Community College, where he has also taught anthropology and sociology. He was also visiting professor of American studies at the Johannes

Gutenberg University Mainz. He received the national AAA/McGraw-Hill Award for Excellence in Undergraduate Teaching of Anthropology from the American Anthropological Association in 2005 and the statewide Hayward Award for Excellence in Education from the California Community Colleges in 2003. He is the author/editor of seven books, including *The Immersive Worlds Handbook*, and he has consulted for the themed entertainment industry. Scott has been interviewed by *To the Best of Our Knowledge*, the *Canadian Broadcasting Company*, *The Independent*, *The New York Times*, *The Washington Post*, *Slate*, *The Chicago Tribune*, *The Financial Times* (of London), *The Daily Beast*, *Huffington Post UK*, *Atlas Obscura*, *Skift*, and *Caravan*, and was part of the documentary film, *The Nature of Existence*.

Markie McBrayer is an assistant professor of political science at the University of Idaho. She received her PhD from the University of Houston. Prior to earning her doctorate, she received her master's in urban and environmental policy and planning from Tufts University, and subsequently worked as a transportation consultant. She studies American politics and policy, with a focus on political institutions, representation, and social inequality in the American context, particularly at the local level. Her work has appeared in *Policy Studies Journal*, *Politics, Policy, and Statistics*, and *Urban Affairs Review*.

Joe R. McBride is professor emeritus of landscape architecture and forest ecology at the University of California, Berkeley. He received his BS in forestry from the University of Montana and MS (forestry) and PhD (botany) from the University of California, Berkeley. His teaching was focused on forest ecology, the ecology of California landscape regions, urban forestry, and the history of American forests. His research ranged from studies of the ecology of California forests to various aspects of urban forestry.

Dr. Kerri Jean Ormerod is an assistant professor in the Department of Geography and Cooperative Extension at the University of Nevada, Reno. She is a social scientist with primary interests in western water law, sociocultural components of risk perception, and Q methodology. Her research typically explores how science, law, and technology interact and adapt to changing social values regarding water governance in the western United States. She also leads Cooperative Extension's Living with Drought program in Nevada.

Camila Pombo, born and raised in Mexico City, is an undergraduate student at the University of Houston, pursuing her Bachelor of Science degree in economics with a minor in energy and sustainability. Before attending her current institution, she went to Lone Star College, CyFair, where she had her first research experience in the Honors College program. Throughout her undergraduate career, she has held student leadership positions in several student organizations and programs, including Women in STEM, National Model United Nations, and, most recently, the Economic Society at the University of Houston. In 2018, she was part of the Calthorpe Project in London, England, where her interest in sustainable food practices and urban gardening expanded. After obtaining her bachelor's degree, she plans to pursue a Ph.D. career path in the area of Economics.

Suzanne Roberts is the author of the travel essay collection *Bad Tourist: Misadventures in Love and Travel* (October 2020) and the memoir *Almost Somewhere: Twenty-Eight Days on the John Muir Trail* (winner of the National Outdoor Book Award), as well as four books of poetry. Named "The Next Great Travel Writer" by National Geographic's Traveler, Suzanne's work has been listed as notable in Best American Essays and included in *The Best Women's Travel Writing*. Her work has appeared in *The New York Times*, CNN, Creative Nonfiction, Brevity, The Rumpus, Hippocampus, The Normal School, River Teeth, and elsewhere. She holds a doctorate in Literature and the Environment from the University of Nevada-Reno and currently teaches for the low residency MFA program in creative writing at Sierra Nevada University. For more information, please visit her website: www.suzanneroberts.net or follow her on Instagram: @SuzanneRoberts28

Juneko J. Robinson is an independent scholar and practicing attorney in immigration law. She received her MA and PhD in philosophy, as well as her Juris Doctor degree in international human rights law from the State University of New York at Buffalo where she was an Arthur A. Schomburg Fellow. Her research interests include existentialism, the philosophy of popular and contemporary political culture, the politics of aesthetics and fashion, and the continental philosophy of law. Prior to this, her most recent work focused on the 1969 Hard Hat Riots and the politics of competing temporalities in 1960s vernacular fashion.

Dr. Loretta Singletary is a professor of economics and an Interdisciplinary Outreach Liaison with Cooperative Extension at the University of Nevada, Reno, with a PhD in applied economics. She has twenty-eight years of experience with land grant universities developing, implementing, evaluating, and administering collaborative research programs. Her programs contribute to the science of social learning and problem-solving by convening interdisciplinary science teams with local policy-makers and resource managers to jointly research complex and often contentious natural resource issues.

Current programs include community climate resiliency; Native American reservation land tenure, water, and economies; and sustaining water resources through a systems approach to foods production.

Mark Thorsby is a professor of philosophy at Lone Star College, CyFair outside Houston, Texas. He has written numerous articles on environmental ethics, eco-phenomenology, architecture, media, and the problem of intersubjectivity. He received his PhD from the New School for social research for his work on Husserlian phenomenology, intersubjectivity, and climate change.

www.ingramcontent.com/pod-product-compliance
Lightning Source LLC
Chambersburg PA
CBHW050637280326
41932CB00015B/2688
* 9 7 8 1 7 9 3 6 1 7 6 5 1 *